SUPER
SPACE

SUPER
SPACE

AUTHOR CLIVE GIFFORD
CONSULTANT DR JACQUELINE MITTON

CONTENTS

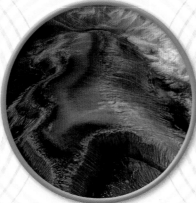

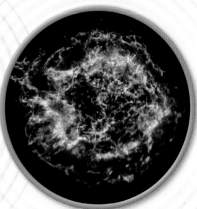

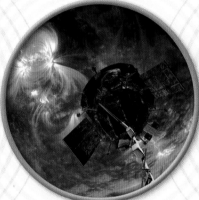

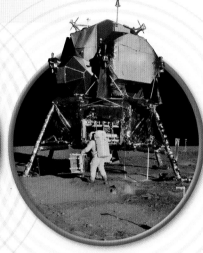

Penguin Random House

DK DELHI
Senior Editor Sreshtha Bhattacharya
Senior Art Editor Vikas Chauhan
Editor Isha Sharma
Art Editor Heena Sharma
Assistant Editor Kathakali Banerjee
Assistant Art Editor Tanisha Mandal
3D Illustrators Mahipal Singh, Nain Rawat
Jacket Designer Suhita Dharamjit
Jackets Editorial Coordinator Priyanka Sharma
Senior DTP Designers Shanker Prasad, Harish Aggarwal
DTP Designer Ashok Kumar
Picture Researcher Nishwan Rasool
Managing Jackets Editor Saloni Singh
Pre-Production Manager Balwant Singh
Production Manager Pankaj Sharma
Picture Research Manager Taiyaba Khatoon
Managing Editor Kingshuk Ghoshal
Managing Art Editor Govind Mittal

DK LONDON
Senior Editors Shaila Brown, Ben Morgan
Senior Art Editor Smiljka Surla
Jacket Designer Stephanie Cheng Hui Tan
Jacket Editor Emma Dawson
Jacket Design Development Manager Sophia MTT
Producer, Pre-Production Gillian Reid
Producer Meskerem Berhane
Managing Editor Lisa Gillespie
Managing Art Editor Owen Peyton Jones
Publisher Andrew Macintyre
Associate Publishing Director Liz Wheeler
Art Director Karen Self
Design Director Phil Ormerod
Publishing Director Jonathan Metcalf

Illustrators Peter Bull, Arran Lewis, Brendan McCaffrey

First published in Great Britain in 2019 by
Dorling Kindersley Limited,
80 Strand, London, WC2R 0RL

A CIP catalogue record for this book is available
from the British Library.
ISBN: 978-0-2413-4344-9

Printed and bound in China

A WORLD OF IDEAS:
SEE ALL THERE IS TO KNOW

www.dk.com

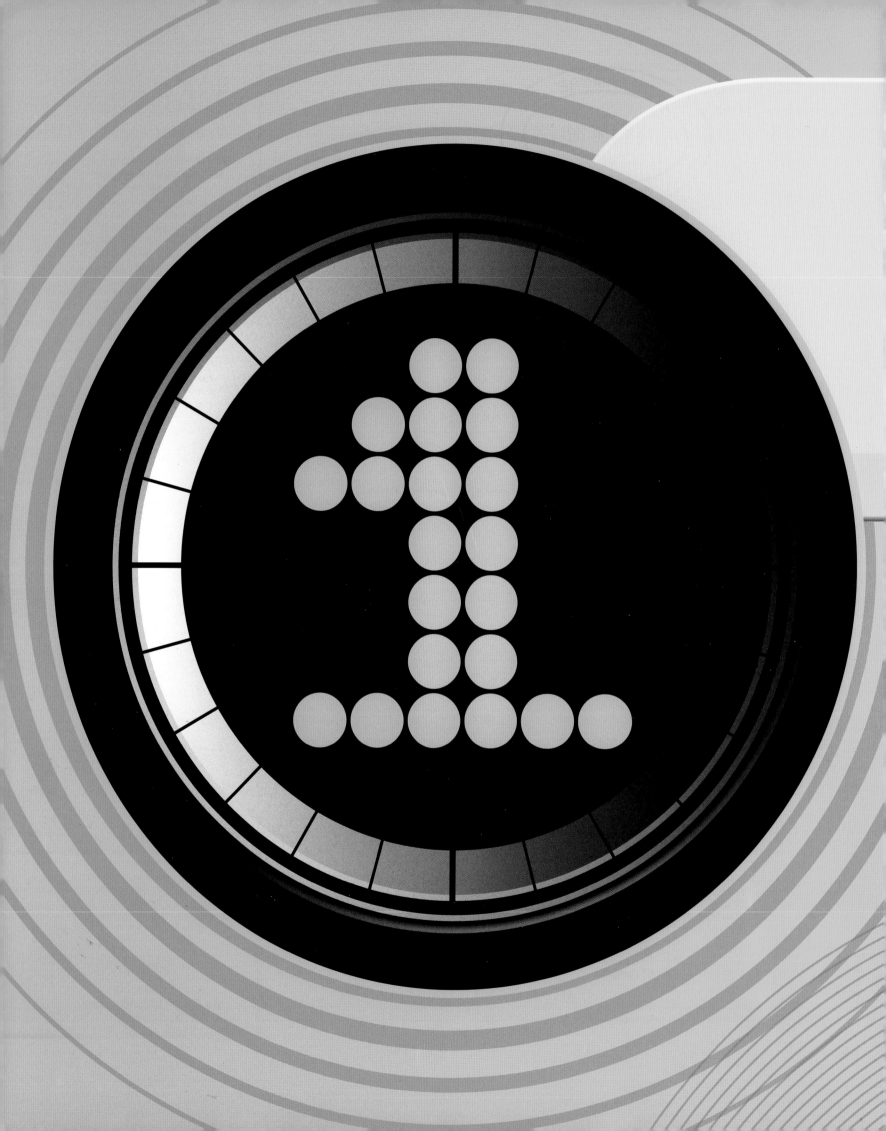

THE UNIVERSE

We are part of the Universe, which includes everything around – from what we already know to what we have yet to learn. The Universe has been expanding since it exploded into existence 13.8 billion years ago. Starting out smaller than a single atom, the Universe is now a vast source of cosmic wonders.

EVOLUTION OF THE UNIVERSE

The Universe is everything that exists, from the smallest particles of dust to giant galaxies of stars. Scientists believe the Universe began 13.8 billion years ago with an event called the Big Bang. This was not an explosion of matter through space but an expansion of space itself, causing all the matter and energy in the Universe to spread out. This process, called inflation, has continued ever since.

13.8 BILLION YEARS AGO

The Big Bang

When the Universe began, all the matter and energy that would ever exist was squeezed into a phenomenally tiny space, smaller than a single atom. In a fraction of a second it expanded at an incredibly rapid rate, ballooning to a size bigger than a planet.

1 SECOND LATER

Matter formed

Within the first second, the Universe expanded to about 100 billion km (60 billion miles) wide. It was full of energy and incredibly hot, but as it expanded it began to cool slightly, and some of the energy formed particles of matter. These would eventually form protons and neutrons – the building blocks of atoms.

400,000 YEARS LATER

First atoms

The Universe cooled down enough for the protons and neutrons to form the first complete atoms of hydrogen and helium, which would later give rise to all the chemical elements we know today. Over millions of years, clouds of hydrogen and helium gas filled the Universe.

200 MILLION YEARS LATER

Stars and galaxies

Gravity pulled on the gas clouds, drawing them together in places to form denser pockets. At the centre of these pockets, the gas eventually became hot enough to trigger nuclear reactions, igniting the first stars. Newborn stars clustered together to make galaxies. The first galaxies were small, but they gradually merged into giant whirlpools of billions of stars.

4.6 BILLION YEARS AGO

The Solar System

Our Sun formed from a cloud of gas and dust left behind by dead stars. Not all this matter was absorbed by the new star – some of it formed a swirling disc around the Sun. Over time, the particles of leftover matter clumped together to form the planets, moons, asteroids, and comets of our Solar System.

4.3 BILLION YEARS AGO

Life began

Earth was just the right distance from the Sun for water to collect as a liquid on its surface and for a thick atmosphere of gas to form around the planet. These conditions allowed the first forms of life to appear in the sea. Over billions of years they evolved into the great diversity of species alive today.

THE SCALE OF SPACE

The Andromeda Galaxy as seen from Earth

LIGHT YEARS

Large distances in the Universe are often measured in light years. The distance light travels through space in a year is about 9.5 trillion km (6 trillion miles). So one light year is a huge distance. When stargazers look at the Andromeda Galaxy, which lies 2.5 million light years away, they are peering back in time, seeing the galaxy when light left it 2.5 million years ago.

Space seems enormous to us, but everything we can ever possibly see is only a fraction of its actual size. Our home planet is a small rotating rock in the expanding Universe. It orbits a star we call the Sun as part of the Solar System. The Sun is located in the Milky Way galaxy, together with billions of other stars. Beyond this are billions more galaxies of different shapes and sizes extending throughout the Universe. The totality of space, with its vast structures, might be infinite.

The Solar System is located in the Orion Arm of the Milky Way.

The Sun is the biggest object in our Solar System.

Blue planet

From space, Earth looks blue because more than two-thirds of the surface is covered in water. It is the largest rocky planet in the Solar System and the only place in the Universe where life is known to exist.

Solar System

Our neighbourhood in space is called the Solar System. It includes millions of comets and asteroids as well as Earth and seven other planets. Together they orbit a central star, the Sun.

Local stars

Beyond our Solar System is the stellar neighbourhood. There are at least 2,000 stars found within 50 light years from the Sun. But they are just a tiny fraction of the billions of stars in our galaxy, the Milky Way.

Milky Way

The Milky Way is an enormous disc-shaped galaxy. The planets of the Solar System lie within this galaxy together with more than 200 billion stars. Spiral arms of gas, dust, and stars wind outwards from the centre.

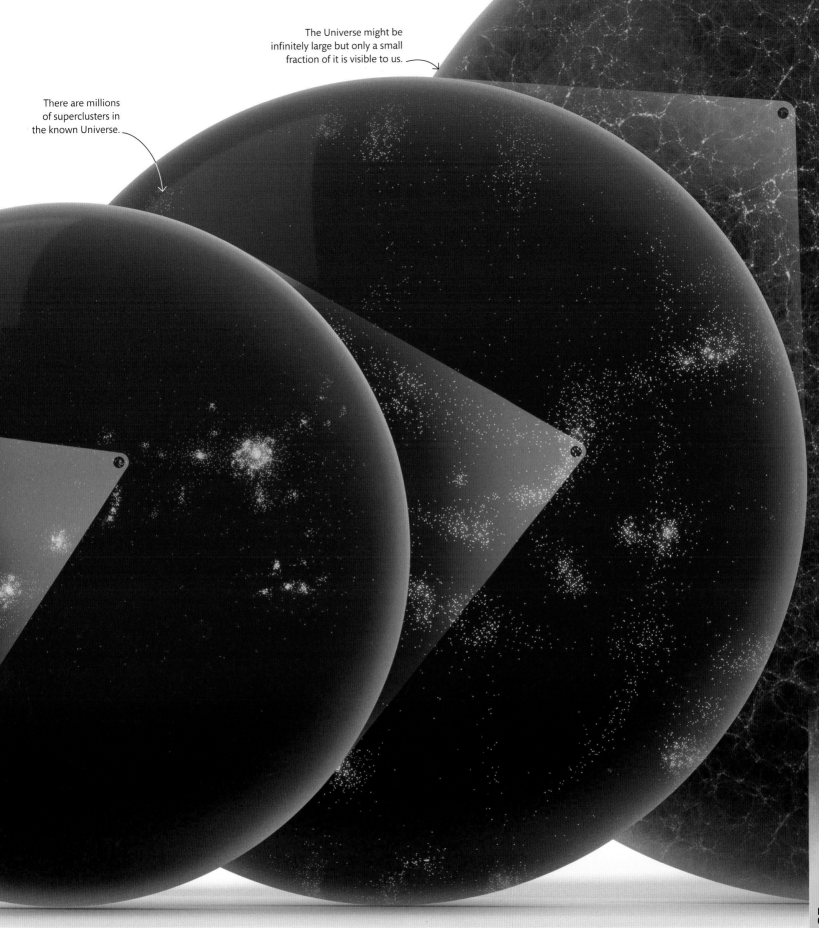

There are millions of superclusters in the known Universe.

The Universe might be infinitely large but only a small fraction of it is visible to us.

Local Group

Galaxies cluster together, held by gravity. About 50 galaxies, including the Milky Way, form a cluster called the Local Group. The Andromeda Galaxy and the Large and Small Magellanic Clouds are members of the Local Group visible without a telescope.

Superclusters

Clusters of galaxies link to create vast superclusters. The Local Group, for example, is part of the Virgo Supercluster. Scientists think that there are at least 100 galaxy groups within the Virgo Supercluster.

Observable Universe

Superclusters are grouped into thread-like structures called filaments. These are thought to be the largest structures in space. Vast, empty spaces separate the filaments. There are superclusters throughout space for as far as we can see.

GALAXIES

Our Milky Way is one of more than 125 billion galaxies in the Universe. These colossal collections of stars, dust, and gas spin through space bound together by gravity. The smallest are dwarf galaxies, each containing fewer than a million stars, while the biggest are giants with hundreds of billions. Altogether, they fill only two-millionths of the Universe due to the vast empty spaces between them. Galaxies first formed billions of years ago and today astronomers classify them based on their shapes in the sky.

"IC 1101, an elliptical galaxy, is one of the largest known in the Universe."

BARRED SPIRAL GALAXIES

Our Milky Way is an example of a barred spiral galaxy. These galaxies feature the long curving arms of a spiral galaxy together with a bar containing bright, young stars in the centre. The two spiral arms of the galaxy NGC 1300 are shown here curving away from the starry bar.

ELLIPTICAL GALAXIES

Elliptical galaxies are swarms of mainly old stars with very little gas and dust. Shaped like a round or squashed ball, they range in size from dwarfs containing tens of millions of stars to the largest known giant galaxies with trillions of stars, such as NGC 1132 shown here.

SPIRAL GALAXIES

About 70 per cent of the galaxies closest to us are spinning spirals, known for their spectacular shapes. Spiral galaxies, such as M81, consist of a giant disc with a central bulge and long, curving arms. The central bulge contains the older stars, while the swirling arms contain the youngest stars.

LENTICULAR GALAXIES

With shapes between spiral and elliptical galaxies, lenticulars have a disc and a central bulge, rather like a lens, but no spiral arms. Many of these galaxies have used up all their gas to create new stars but some, like NGC 6861 shown here, still contain lanes of dust.

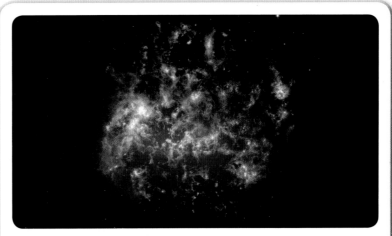

IRREGULAR GALAXIES

Galaxies that have no recognizable shape are called irregular galaxies. Full of gas and dust, they are home to lots of young stars and bright nebulas. They are usually small, like the Large Magellanic Cloud seen here, with a diameter only one-tenth of the Milky Way. Irregulars may have lost their original shape after collisions with other galaxies.

INTERACTING GALAXIES

Most galaxies are spread far apart, but it is possible for two galaxies to drift so close that their gravity disrupts and distorts each other. The spiral arms of UGC 1810, shown here, have been pulled out of shape by the gravity of its companion, UGC 1813. This interaction produces clouds of gas and triggers the formation of new stars.

UGC 1813 is one-fifth the size of its larger companion galaxy UGC 1810.

BLACK HOLES

Almost all galaxies have supermassive black holes at their centres. While the black holes themselves are invisible, astronomers can observe the whirling disc of surrounding materials being pulled into them. The gravity around a black hole is so strong that not even light can escape from it.

TYPES OF STAR

In the night sky, all stars appear to us as tiny sparkling lights. However, each one is different and has its own characteristics. Stars range in size from dwarfs smaller than Earth to supergiants 1,000 times bigger than the Sun. The temperatures, colours, and brightness of stars also vary. All these characteristics, as well as a star's lifetime, are determined by its mass and its age. Astronomers use them to classify stars into types.

SIZING UP

This artist's impression compares the sizes of stars of different types and temperatures. The coolest stars shine red or orange, while the hottest stars appear blue or white. More massive stars are hotter and brighter, but they have shorter lifespans.

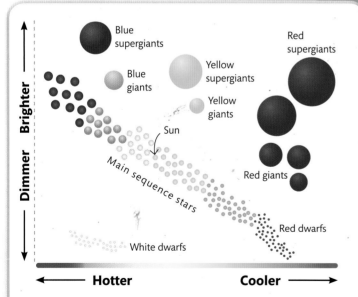

CLASSIFYING STARS

On a diagram of brightness against temperature, stars such as the Sun (a yellow dwarf), which are in mid-life and burning hydrogen in their cores, form a band called the main sequence. Giants and supergiants group together at the top while tiny white dwarfs lie under the main sequence.

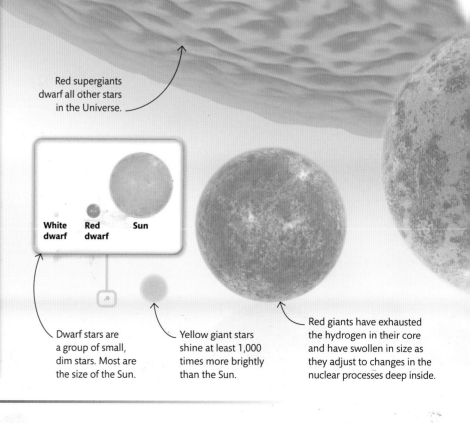

Red supergiants dwarf all other stars in the Universe.

Dwarf stars are a group of small, dim stars. Most are the size of the Sun.

Yellow giant stars shine at least 1,000 times more brightly than the Sun.

Red giants have exhausted the hydrogen in their core and have swollen in size as they adjust to changes in the nuclear processes deep inside.

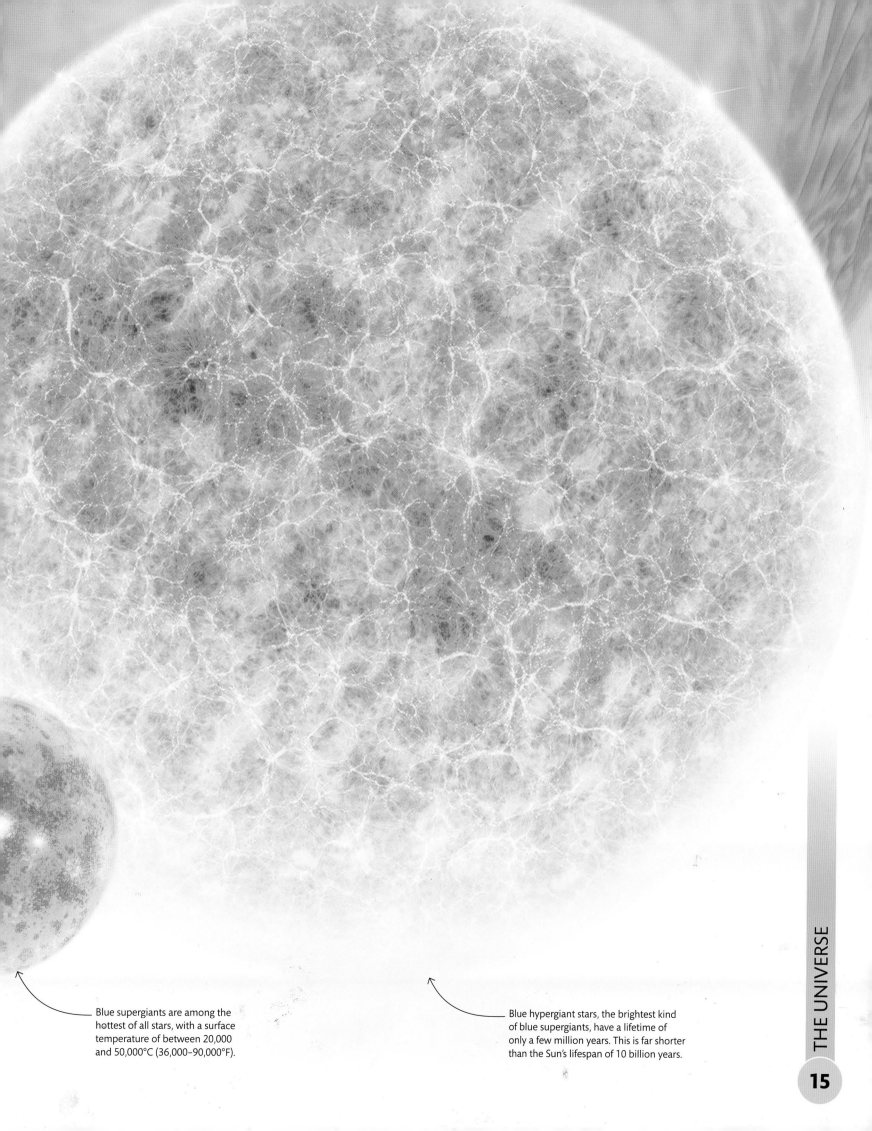

Blue supergiants are among the hottest of all stars, with a surface temperature of between 20,000 and 50,000°C (36,000–90,000°F).

Blue hypergiant stars, the brightest kind of blue supergiants, have a lifetime of only a few million years. This is far shorter than the Sun's lifespan of 10 billion years.

LIVES OF STARS

Stars form throughout the Universe and may last for billions of years. Small, medium, and massive stars have distinctive life cycles and evolve at different rates. All types of stars are fuelled by nuclear reactions in their cores, which produce a lot of energy, including heat and light. When its fuel runs out, a star may die suddenly or gradually. The way the star dies depends on how massive it is.

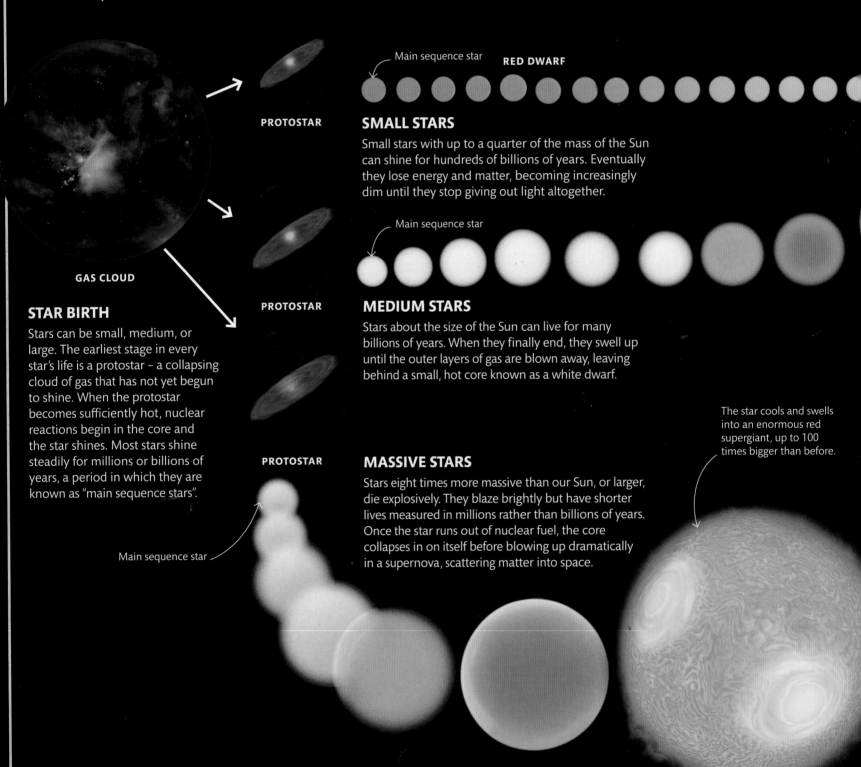

GAS CLOUD

PROTOSTAR

PROTOSTAR

PROTOSTAR

Main sequence star

Main sequence star — RED DWARF

Main sequence star

STAR BIRTH

Stars can be small, medium, or large. The earliest stage in every star's life is a protostar – a collapsing cloud of gas that has not yet begun to shine. When the protostar becomes sufficiently hot, nuclear reactions begin in the core and the star shines. Most stars shine steadily for millions or billions of years, a period in which they are known as "main sequence stars".

SMALL STARS

Small stars with up to a quarter of the mass of the Sun can shine for hundreds of billions of years. Eventually they lose energy and matter, becoming increasingly dim until they stop giving out light altogether.

MEDIUM STARS

Stars about the size of the Sun can live for many billions of years. When they finally end, they swell up until the outer layers of gas are blown away, leaving behind a small, hot core known as a white dwarf.

MASSIVE STARS

Stars eight times more massive than our Sun, or larger, die explosively. They blaze brightly but have shorter lives measured in millions rather than billions of years. Once the star runs out of nuclear fuel, the core collapses in on itself before blowing up dramatically in a supernova, scattering matter into space.

The star cools and swells into an enormous red supergiant, up to 100 times bigger than before.

"A single supernova can outshine an entire galaxy full of billions of stars."

STUDYING STARS

In 1925, the British-American astronomer Cecilia Payne-Gaposchkin (1900–1979) found that stars are mainly composed of hydrogen and helium, the two lightest chemical elements. All the other heavier elements, such as iron and oxygen, make up less than two per cent of the mass of stars.

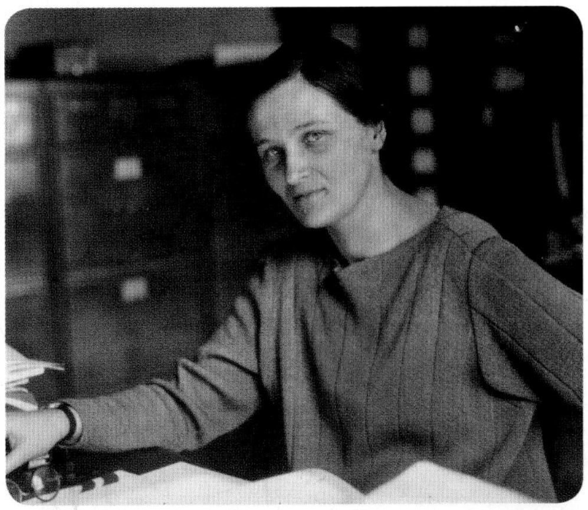

BLUE DWARF

Star's light begins to dim

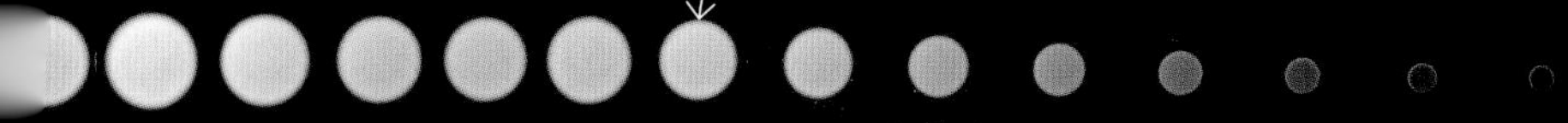

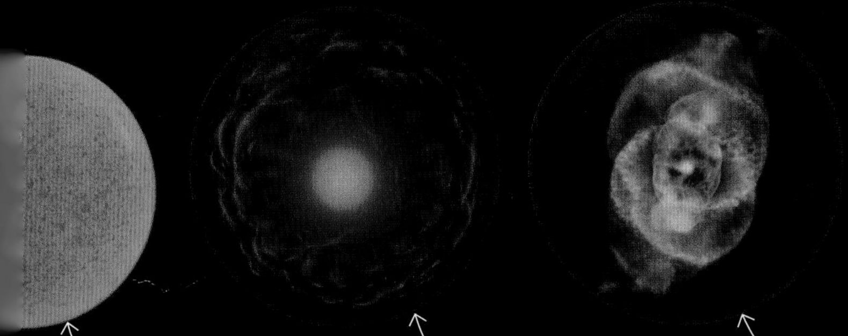

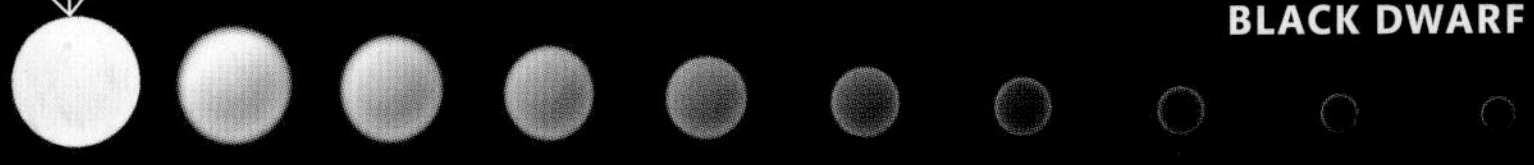

BLACK DWARF

A red giant forms when hydrogen fuel in the core runs out and the star cools and expands.

The star's outer layers disperse into space when its fuel runs out.

The dying star is surrounded by a glowing cloud of gas and dust, which forms a planetary nebula.

The remaining, glowing core is called a white dwarf star. It gradually dims over billions of years.

BLACK DWARF

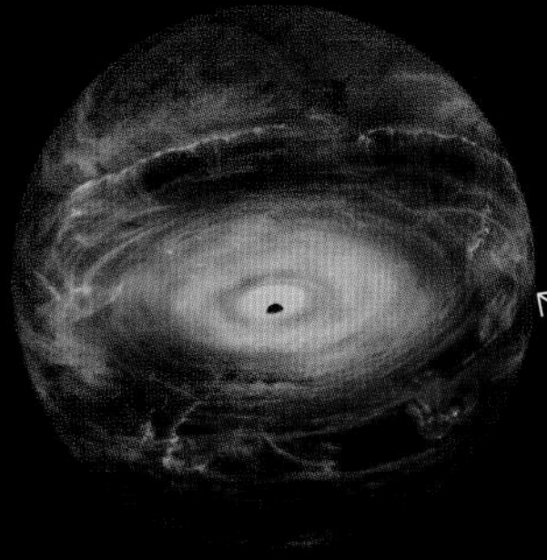

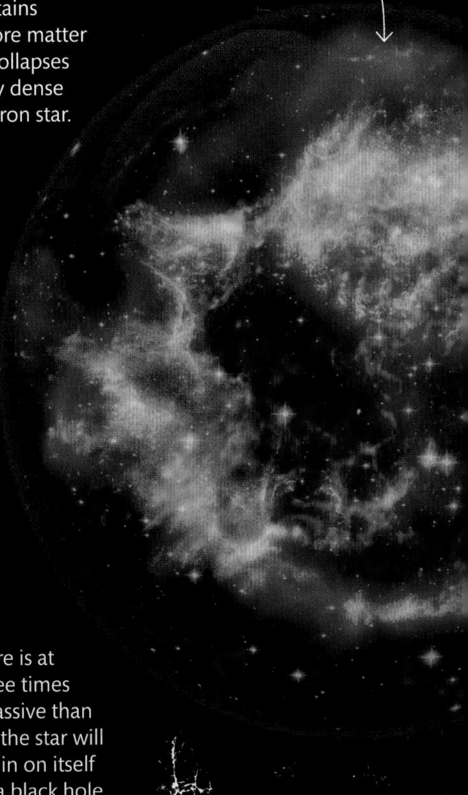

The weakening core can no longer support the star's enormous mass. It collapses and then rebounds in a supernova explosion, blasting off its outer layers with phenomenal force.

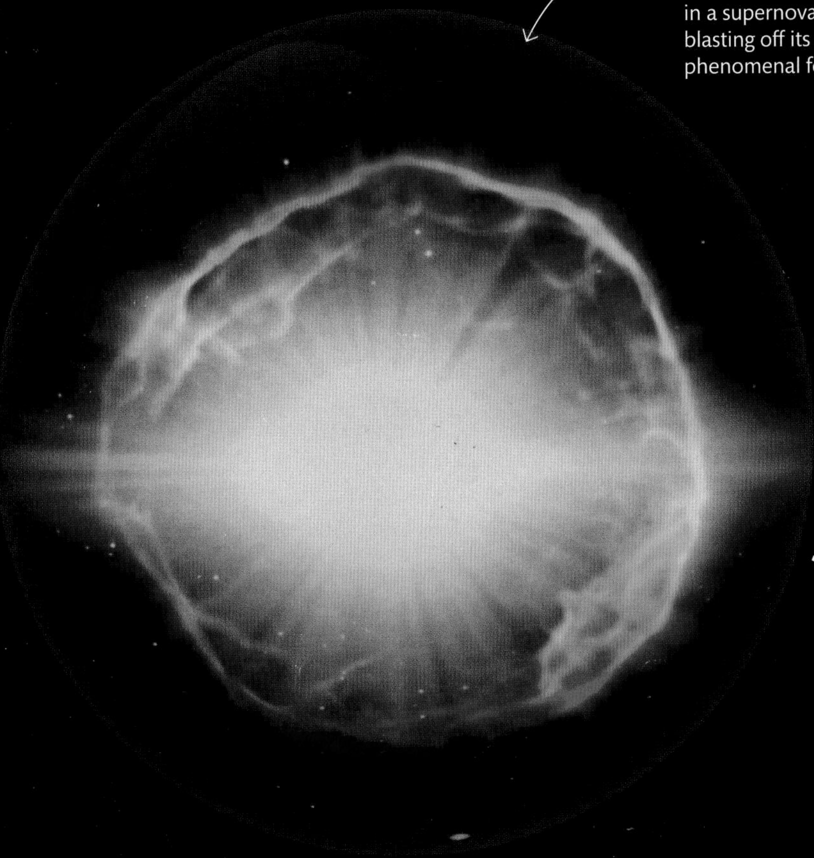

If the core left behind after a supernova contains 1.4 to 3 times more matter than the Sun, it collapses into an incredibly dense star called a neutron star.

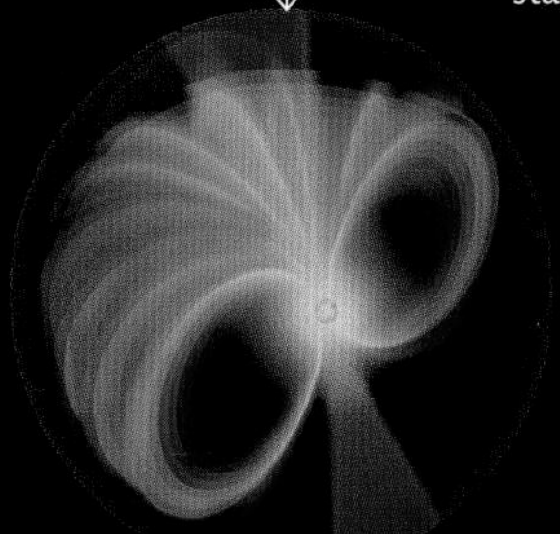

If the core is at least three times more massive than the Sun, the star will collapse in on itself to form a black hole.

Supernovas produce vast clouds of gas and dust that disperse into space and may go into new stars.

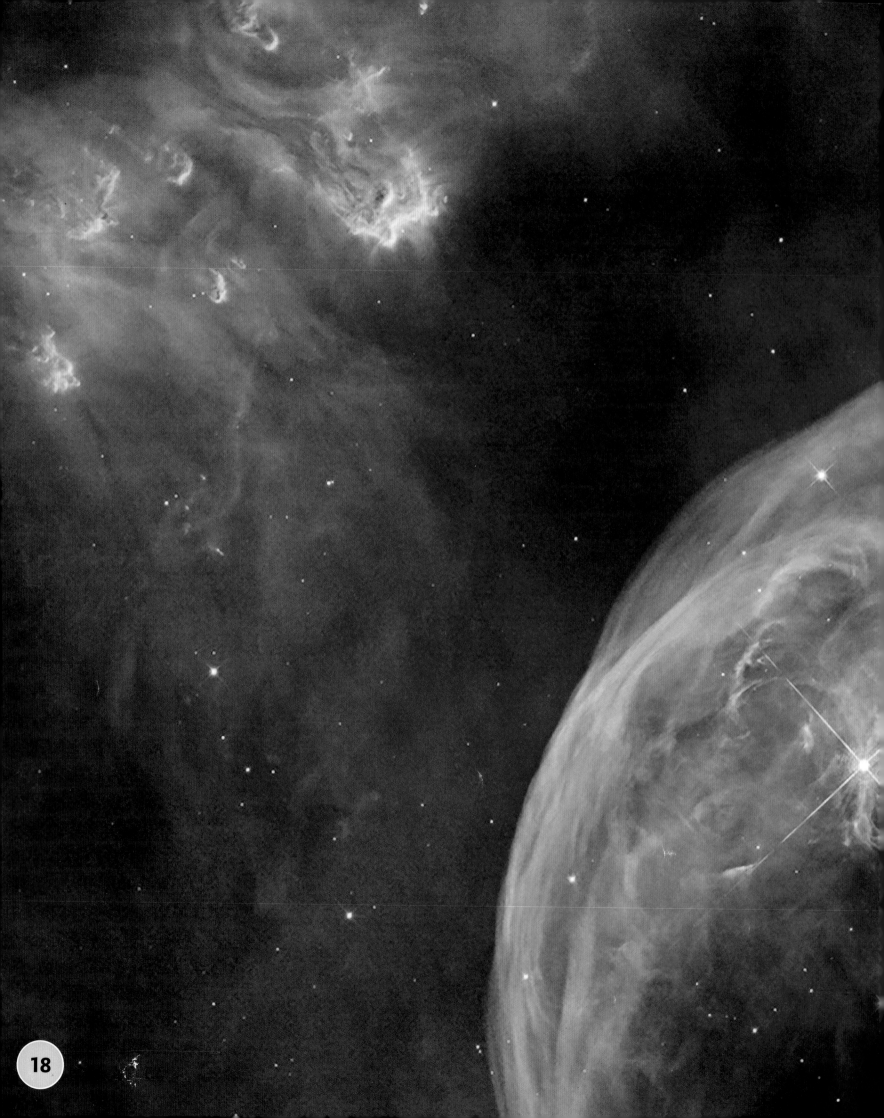

SPACE BALLOON

This amazing image of the Bubble Nebula was captured by the Hubble Space Telescope. It shows a huge balloon-like cloud of gas and dust, which is being blown into space by a giant star 45 times more massive than the Sun. Newborn stars scattered within the nebula are illuminating the dusty cloud.

MAKING PLANETS

The Solar System formed more than eight billion years after the Big Bang. A sudden burst of star formation inside a gas cloud triggered the formation of the Sun. The surrounding ring of gas and dust became the planets, dwarf planets, moons, asteroids, and comets. This planet-making process began 4.5 billion years ago and took another 700 million years to complete, making the complex structure we are part of today. The last leftovers of the debris that generated our Solar System still fall to Earth as space rocks called meteorites.

> "Today the Sun contains 99.8% of all the material in the Solar System."

GAS AND ICE GIANTS

In the coldest outer reaches of the disc, icy particles began to bind together. Some became so big that their gravity pulled in hydrogen, helium, and other gases. These formed the gas giants Jupiter and Saturn, and the ice giants Uranus and Neptune.

BIRTH OF THE SOLAR SYSTEM

The Sun was born when a giant cloud of gas and dust collapsed under its own gravity. All the material pulled into the middle made up the Sun, while some particles were left in a spinning disc around the new star. Over millions of years, clashes and collisions would turn this whirling gas and dust into the Solar System, including our home planet Earth.

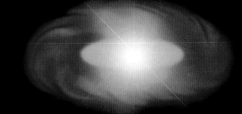

Shrinking gas

A pocket of gas inside a huge cloud in space was pulled together by its own gravity. As it shrank, it started to spin and heat up in the centre.

Hot centre

Nuclear reactions started up in the hot centre, which began glowing. The Sun formed here, surrounded by a spinning disc.

SOLAR NEBULA

As the young Sun formed inside a huge spinning cloud of gas and dust, a disc of material developed around its equator. This disc is now known as the solar nebula. The Sun heated up, causing most of the gas to expand away into space, leaving behind hot rock and metals in the centre. Away from the Sun, the most distant areas of the disc were cold enough for water and gases to freeze solid.

LUNAR FORMATION

When the young Earth was still molten lava, it was struck by a rock the size of Mars. The collision threw part of Earth's crust into space, forming a cloud of debris around our planet. Over time, this material clumped together to form a neighbouring rocky body called the Moon.

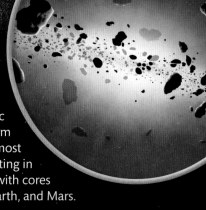

ROCKY PLANETS

The hot centre of the spinning disc contained rocky and metallic particles, which joined together to form protoplanets. Their gravity swept most other material out of the way, resulting in the formation of four rocky planets with cores of iron – Mercury, Venus, Earth, and Mars.

Planetesimals

As the disc rotated, particles clumped together into billions of small bodies, called planetesimals.

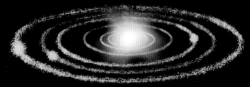

Planets

Collisions between planetesimals resulted in a small number of larger bodies. These became the eight planets of the Solar System.

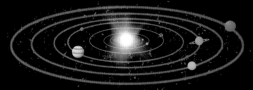

The Solar System

The orbits of the giant planets altered over time until they settled down at their current distances from the Sun about 3.9 billion years ago.

THE SOLAR SYSTEM

Eight planets, five dwarf planets, over 500 moons, and millions of asteroids and comets make up our Solar System around our nearest star, the Sun. Held in place by the Sun's gravity, most of these bodies travel around the Sun on paths known as orbits. The farther away from the Sun an object lies, the longer its orbit takes. Our Solar System is just one of billions of star systems in the Universe.

Comets
These chunks of dusty ice heat up as they approach the Sun, releasing dust and gas, which stream behind the comet in long tails.

Neptune
This distant planet is 30 times farther away from the Sun than Earth. Neptune takes 163.7 years to complete one orbit around the Sun.

THE PLANETS

The four rocky worlds nearest to the Sun are called the inner planets. They are much smaller than the four giant outer planets, which are mainly made of gas. More than 150 known moons orbit planets in the Solar System.

Jupiter
Twice as massive as the other seven planets combined, this giant globe of gas has 79 moons orbiting it, the most of any planet.

Uranus
This ice giant travels around the Sun tilted on its side, and is orbited by 27 moons.

Asteroid Belt
Some of the rocky leftovers from the birth of the Solar System occupy a broad region of space between the orbits of Mars and Jupiter. More than 1.1 million asteroids have a diameter of 1 km (¾ mile) or greater.

THE OORT CLOUD

The Oort Cloud is a shell of comets and space debris that surrounds the outer Solar System. Although it hasn't been directly seen, scientists believe it contains more than one trillion mostly small, icy objects. Beginning past the Kuiper Belt, it is thought to stretch into space as far as 100,000 times the distance between Earth and the Sun.

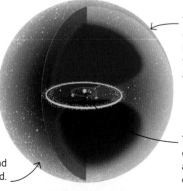

The outer edge of the Oort Cloud may extend more than 1–2 light years away from the Sun.

The Sun and the planets occupy the centre of this area, far from the innermost edge of the Oort Cloud.

Comets that have very long orbits around the Sun come from the Oort Cloud.

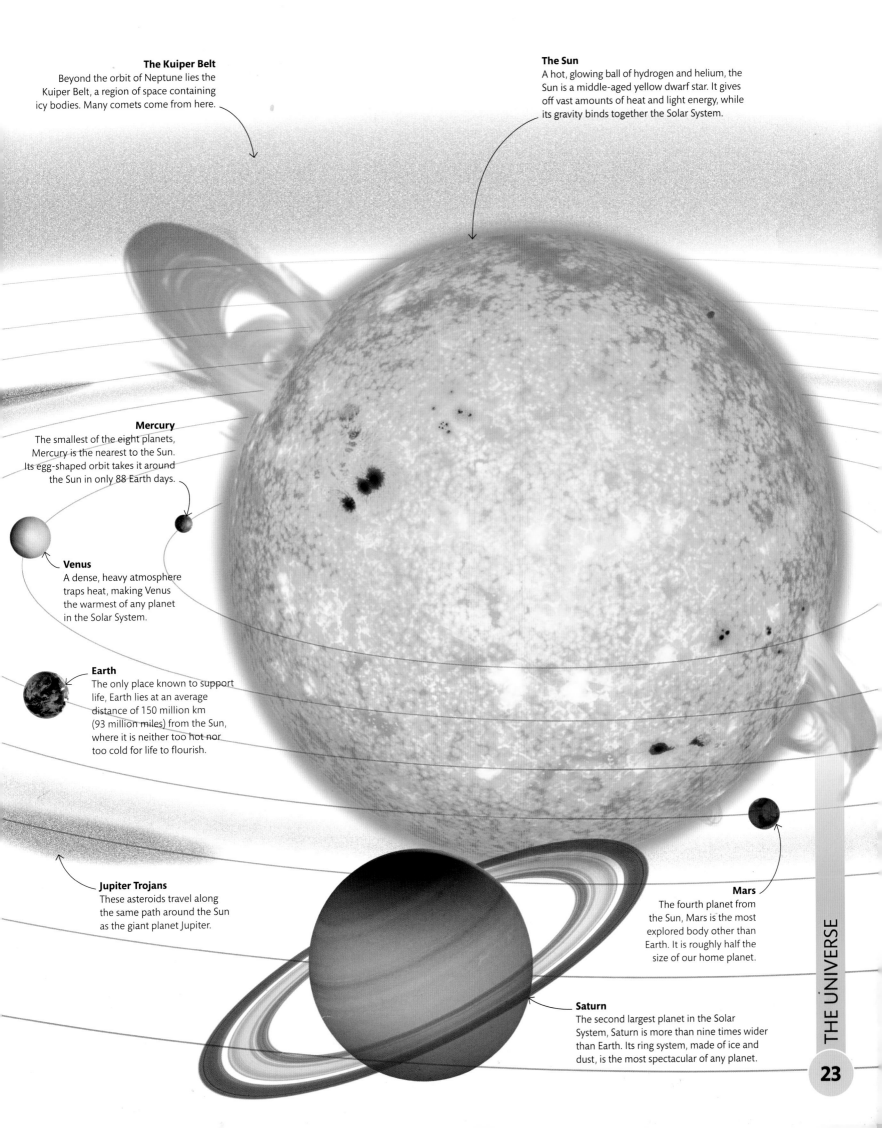

The Kuiper Belt
Beyond the orbit of Neptune lies the
Kuiper Belt, a region of space containing
icy bodies. Many comets come from here.

The Sun
A hot, glowing ball of hydrogen and helium, the
Sun is a middle-aged yellow dwarf star. It gives
off vast amounts of heat and light energy, while
its gravity binds together the Solar System.

Mercury
The smallest of the eight planets,
Mercury is the nearest to the Sun.
Its egg-shaped orbit takes it around
the Sun in only 88 Earth days.

Venus
A dense, heavy atmosphere
traps heat, making Venus
the warmest of any planet
in the Solar System.

Earth
The only place known to support
life, Earth lies at an average
distance of 150 million km
(93 million miles) from the Sun,
where it is neither too hot nor
too cold for life to flourish.

Jupiter Trojans
These asteroids travel along
the same path around the Sun
as the giant planet Jupiter.

Mars
The fourth planet from
the Sun, Mars is the most
explored body other than
Earth. It is roughly half the
size of our home planet.

Saturn
The second largest planet in the Solar
System, Saturn is more than nine times wider
than Earth. Its ring system, made of ice and
dust, is the most spectacular of any planet.

THE UNIVERSE

23

OBSERVING SPACE

Astronomers can see deep into space thanks to the variety of telescopes and observatories available today. Distant objects emit visible light and other forms of energy, which travel through space as waves of different lengths. Traditional optical telescopes gather visible light to create images far clearer than the naked eye can see. Invisible energy can also be collected and studied to build a better understanding of space. All the images here show the Crab Nebula, the remains of an exploded star 6,500 light years from Earth. Each picture shows just one of the different kinds of radiation it emits.

RADIO WAVES

Radio telescopes on Earth use large dish antennae to observe radio waves, which can pass through dust clouds. Both cool gas between stars and extremely hot gas in galaxies and nebulas emit radio waves. This stunning radio map shows filaments of hot gas in the Crab Nebula.

Very Large Array, a radio astronomy observatory

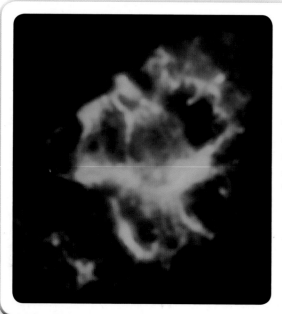

INFRARED RADIATION

Infrared radiation is given off by many objects in space, including the Crab Nebula. With longer wavelengths, it carries less energy than visible light. Some infrared can pass through Earth's atmosphere, but most infrared observations are taken from telescopes in space.

Spitzer Space Telescope, an infrared space observatory

MANY WAVELENGTHS

This spectacular image of the nebula combines the radio (red), infrared (yellow), ultraviolet (blue), and X-ray (purple) images with a Hubble Space Telescope image made with visible light (green). By combining the views from five different telescopes, scientists get a more complete picture with incredible detail that no single image could provide.

ULTRAVIOLET RADIATION

Ultraviolet radiation has shorter wavelengths than visible light and carries more energy. The Crab Nebula's ultraviolet comes from atomic particles spiralling around in a powerful magnetic field. This picture is made up of 75 images taken by an ultraviolet camera on the XMM-Newton X-ray observatory.

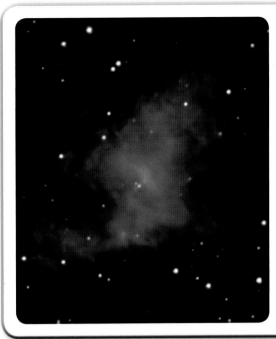

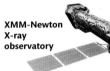

XMM-Newton X-ray observatory

X-RAYS

More powerful than ultraviolet radiation, X-rays are given off by very hot gas and fast-moving atomic particles. This X-ray image of the region around the heart of the Crab Nebula shows jets and rings of high-energy particles blasted away from the pulsar that powers the whole nebula.

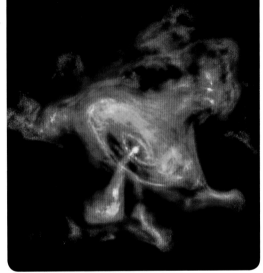

Chandra X-ray observatory

GAMMA RAYS

With the shortest wavelengths, gamma radiation is the most energetic of all. The rays are difficult to capture and gamma-ray images do not have much detail. Most cannot penetrate Earth's atmosphere and must be observed from space. This image shows a gamma-ray flare from the Crab pulsar.

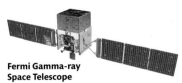

Fermi Gamma-ray Space Telescope

Hubble Space Telescope

SPACE TRAVEL

The era of space travel began in the 20th century with the invention of rockets followed by robotic space explorers and even habitable space stations. Today more than 500 people have visited space, and we have sent uncrewed spacecraft to every planet in the Solar System. This timeline shows some of the key breakthroughs in the history of space exploration.

Goddard's rocket

American engineer Robert Goddard designed the world's first liquid-fuelled rocket. Launched from a ladder-like support structure, the breakthrough invention flew 12 m (41 ft) high for 2.5 seconds.

1926

1942

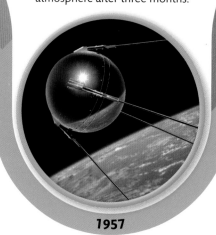

V-2 rocket

German rocket scientist Wernher von Braun built the V-2 rocket as a World War II weapon. The first V-2 rocket travelled 190 km (118 miles). This proved that rockets could one day reach space.

Sputnik 1

This 58 cm (23 in) metal sphere was the first human-made object to orbit Earth. It had a battery-powered transmitter that sent back a "bleep-bleep" signal for 21 days. Sputnik 1 burned up in the atmosphere after three months.

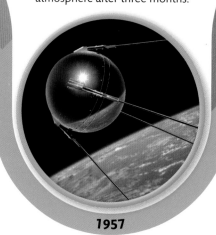

1957

1961

Yuri Gagarin

Russian Yuri Gagarin will always be remembered as the first person in space. On 12 April 1961, he orbited Earth in just under two hours onboard the tiny Vostok 1 spacecraft.

2018

Hayabusa2

The Japanese spacecraft Hayabusa2 reached asteroid Ryugu in October 2018. Three of the four landers it released to explore the surface were the first rovers to move from place to place by hopping.

Philae

On 12 November 2014, the lander Philae became the first probe to land on a comet when it touched down on 67P. It travelled on the Rosetta spacecraft, a journey that took more than 10 years.

2014

1998

International Space Station

About the size of a football pitch, the International Space Station is the largest human-made object to orbit Earth. Building the ISS in space began in 1998, and since 2000 it has been continuously occupied by astronauts.

Sojourner

NASA's Sojourner rover was the first wheeled lander to touch down on Mars. It operated for three months, taking more than 500 photographs and travelling about 100 m (330 ft).

1997

"That's one small step for a man, one giant leap for mankind."

Neil Armstrong

Valentina Tereshkova

Valentina Tereshkova followed Gagarin two years later to become the first woman in space. As the pilot of the Vostok 6 mission, she spent three days in space, orbiting Earth 48 times.

1963

1965

First spacewalk

The Russians were still leading the space race when Aleksei Leonov performed the first spacewalk on 18 March 1965. He was outside Voskhod 2 for 10 minutes, connected by a safety tether.

Apollo 11 landing

The Apollo 11 mission transported the first astronauts to the Moon. The world watched as Americans Neil Armstrong and Edwin Buzz Aldrin took their first steps on the lunar surface on 21 July 1969.

1969

1970

Lunokhod 1

The first rover to explore another world was Russia's Lunokhod 1. In November 1970, the Luna 17 spacecraft carried this remote-controlled rover to the Moon, where it explored its rocky surface for almost ten months.

1986

Mir

Russia's Mir space station was put into Earth orbit in 1986. It was occupied by a crew of three astronauts from 1987–2000. Astronaut Valeri Polyakov spent 437 days on board – the longest time anyone has spent in space.

Columbia

NASA's Space Shuttle Columbia, which flew 28 times, was the first reusable crewed spacecraft. Of the shuttles built, five were launched into space by rockets, and glided back to Earth, landing on a runway.

1982

1976

Viking 1

The first spacecraft to land on Mars was NASA's Viking 1. It spent six years investigating Mars. As well as taking pictures and collecting samples, Viking 1 searched in vain for signs of alien life on the red planet.

Salyut 1

The world's first space station, Salyut 1, was launched by Russia in 1971. Its crew of three lived in the station for 24 days, showing that it was possible for astronauts to live and work in space.

1971

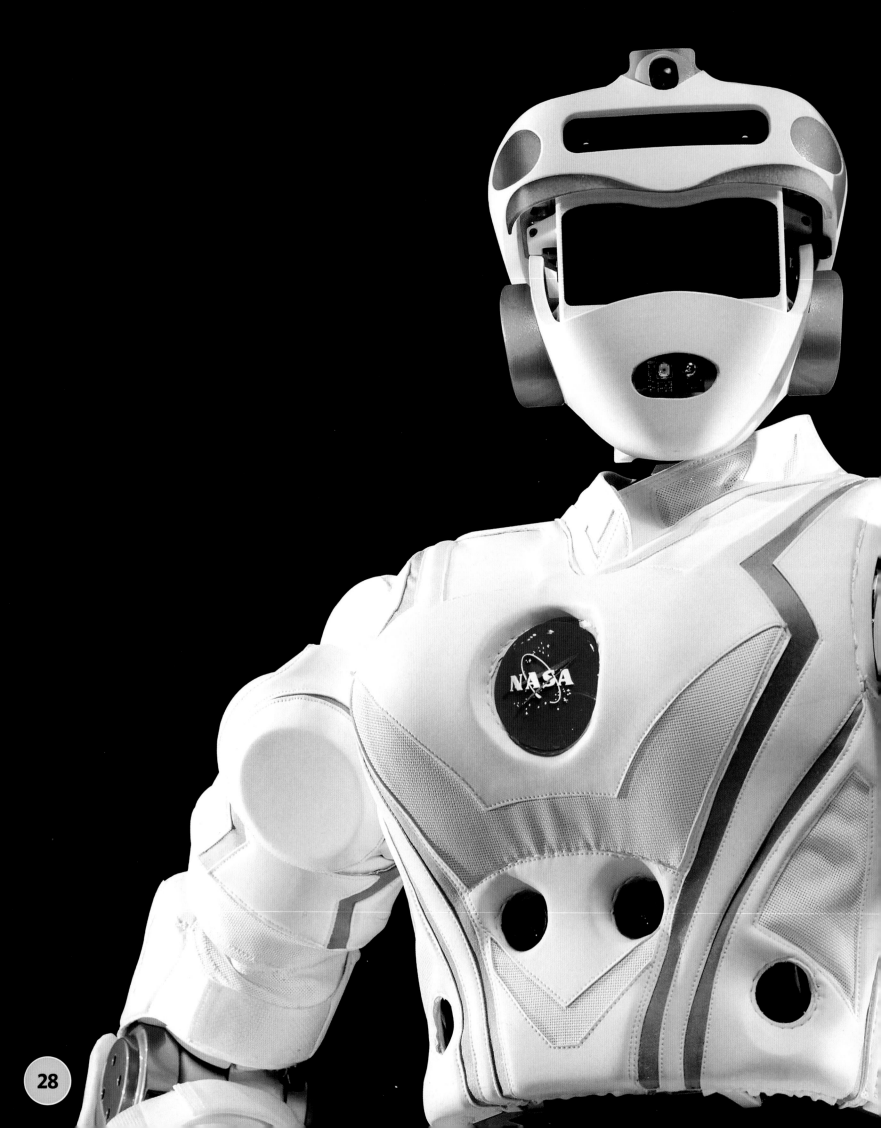

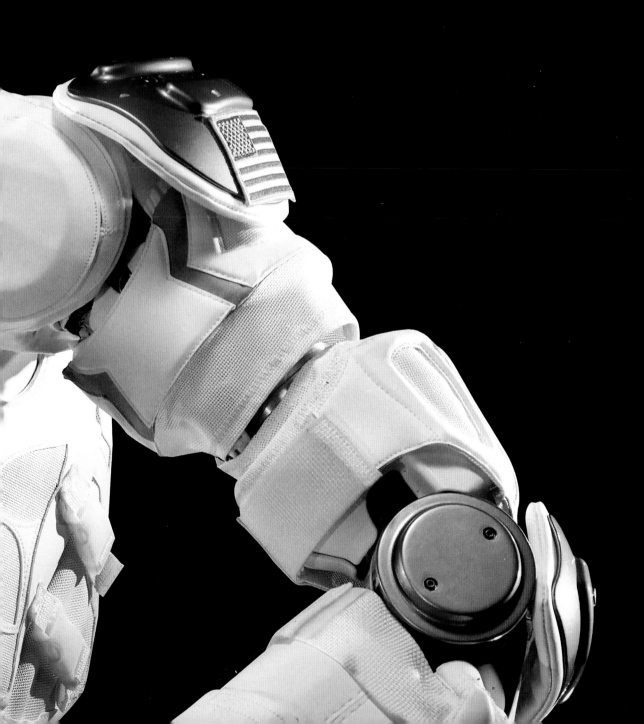

ROBONAUT

R5 is NASA's humanoid robot designed to withstand extreme environments. The robot can move freely, drive, climb, use tools, and stay upright on rough terrain. In future, NASA hopes a similar robot will work alongside astronauts in a space mission to Mars, where it will help build a liveable base for human visitors.

INNER SOLAR SYSTEM

Four rocky worlds lie
at the centre of our
Solar System. Earth, our home
planet and the largest, is one
of them, together with Mercury,
Venus, and Mars. Most space
exploration so far has taken
place within this region.

OUR NEAREST STAR
THE SUN

More than 1 million Earths could fit inside the Sun, the huge, bubbling ball of gas that warms and lights up our planet and the rest of the Solar System. The Sun is made up almost entirely of hydrogen and helium. Nuclear reactions in its core convert hydrogen to helium, releasing energy as light. This light slowly travels out from the core to reach the Sun's surface before spreading out into space.

Inside the Sun

This cutaway image of the Sun shows the fierce nuclear furnace in the star's core and its turbulent visible surface, the photosphere. Loops and jets of hot gas leap from the surface up into the Sun's atmosphere, while radiation escapes the photosphere and travels outwards through the Solar System as heat and light.

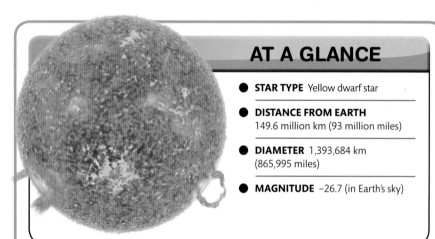

AT A GLANCE

- **STAR TYPE** Yellow dwarf star

- **DISTANCE FROM EARTH**
  149.6 million km (93 million miles)

- **DIAMETER** 1,393,684 km
  (865,995 miles)

- **MAGNITUDE** −26.7 (in Earth's sky)

Bubbles of hot gas expand and rise towards the photosphere through an area called the convective zone.

STORM OF LIGHTS

Apart from heat and light, the Sun also emits streams of high-energy particles that make up the solar wind. Earth's magnetic field mostly shields us from the solar wind, but a strong blast can cause a light display in the atmosphere, called an aurora.

Aurora over Iceland

SOLAR CORONA

The Sun's atmosphere is divided into an inner zone (the chromosphere) and the outer atmosphere, or corona, which extends millions of kilometres into space. Seen here in ultraviolet light, the corona is a halo of gas at a temperature of millions of degrees surrounding the Sun.

The Sun is the nearest star to Earth. It was formed about 4.5 billion years ago, and will shine for another 4.5 billion years, before turning into a red giant that will eventually engulf our planet.

SOLAR WIND

The solar wind can stream through space at speeds of over 750 km/sec (470 miles/sec).

LARGE SUNSPOT

The Great Sunspot of 1947 was large enough to be seen with the naked eye at sunset.

FUEL

The Sun processes over 500 million tonnes of hydrogen every second in its core.

ENERGY

The Sun produces enough energy in one second to power the world for 27,800 years.

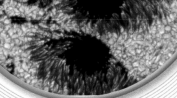

SUNSPOTS

Dark patches sometimes appear on the Sun's surface temporarily. Called sunspots, they look dark because they are up to 2,000°C (3,632°F) cooler than the surrounding areas on the surface.

The Sun's core is where nuclear reactions produce its energy.

The gas below the photosphere is opaque, giving the illusion of a surface.

A bright loop of hot gas, called a prominence, leaps out of the photosphere.

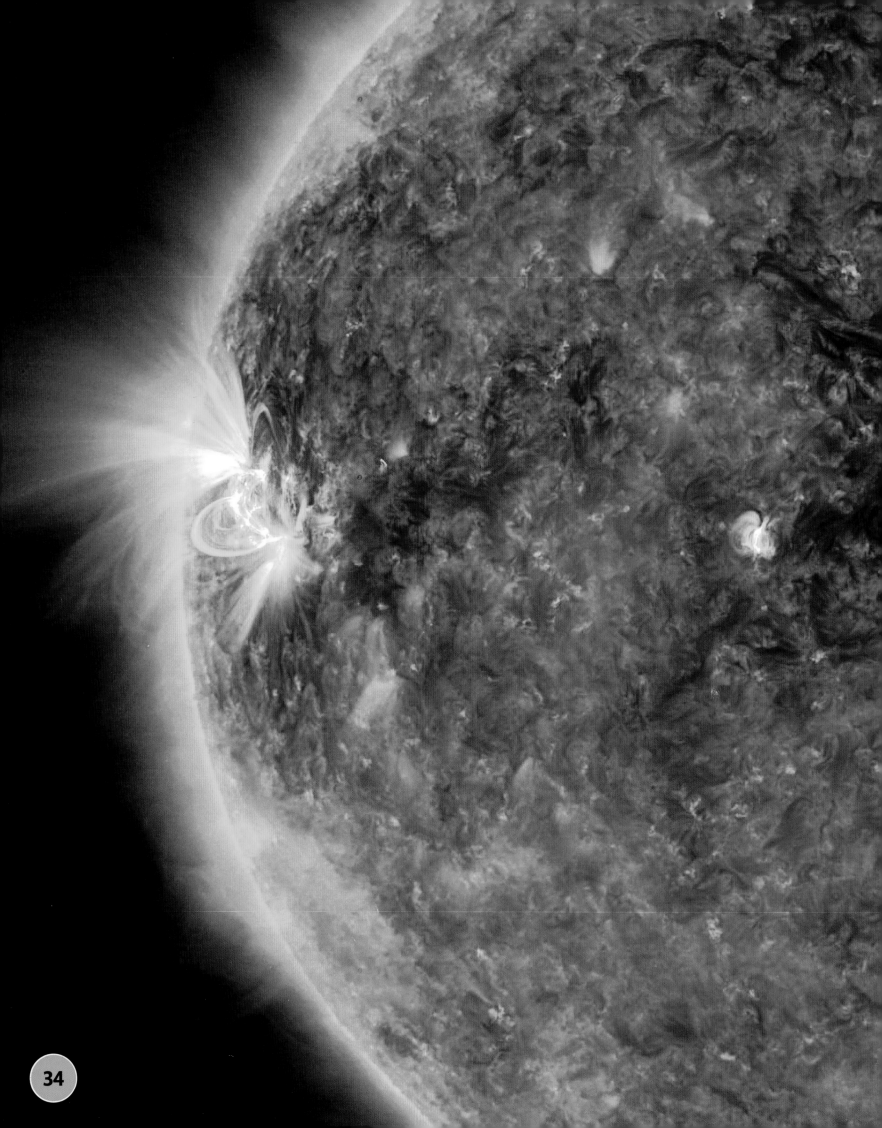

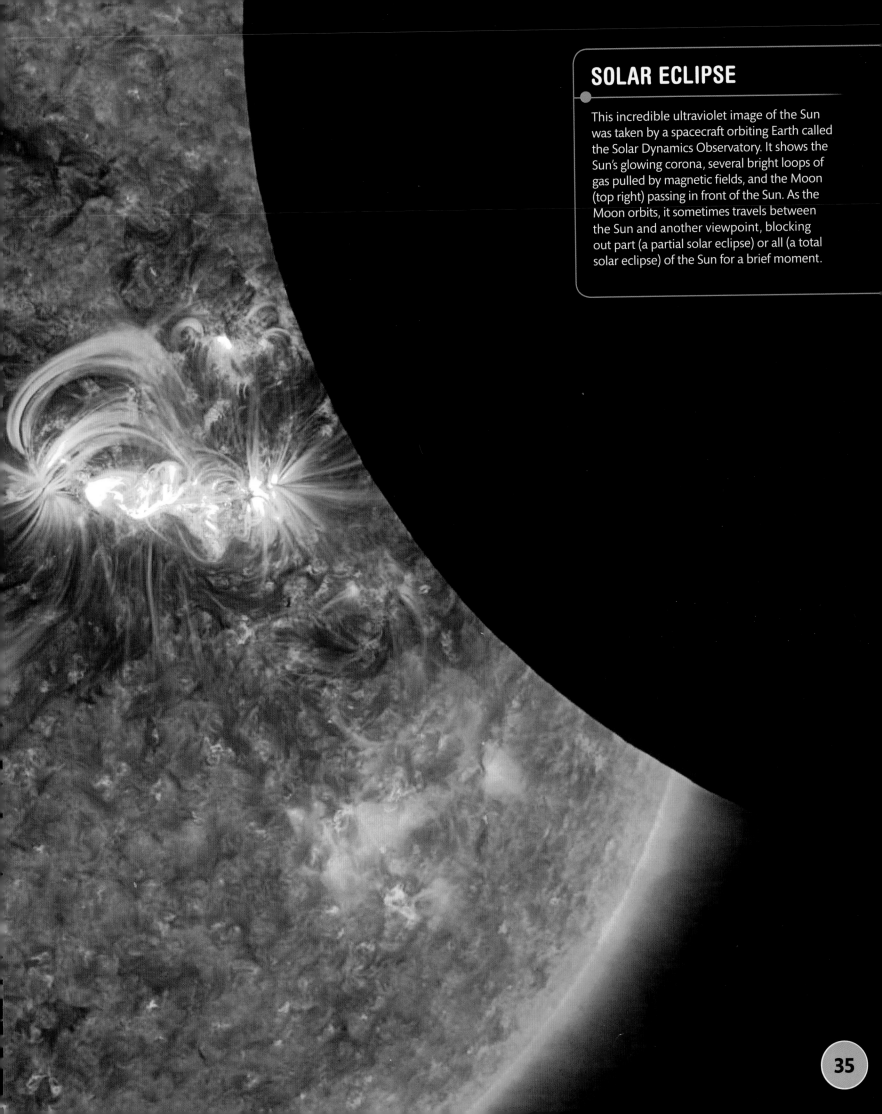

SOLAR ECLIPSE

This incredible ultraviolet image of the Sun
was taken by a spacecraft orbiting Earth called
the Solar Dynamics Observatory. It shows the
Sun's glowing corona, several bright loops of
gas pulled by magnetic fields, and the Moon
(top right) passing in front of the Sun. As the
Moon orbits, it sometimes travels between
the Sun and another viewpoint, blocking
out part (a partial solar eclipse) or all (a total
solar eclipse) of the Sun for a brief moment.

SCARRED PLANET
MERCURY

The smallest and fastest-moving planet in the Solar System is brutally scarred by thousands of craters, where asteroids and meteorites have bombarded its surface. Away from the craters, long cliffs called scarps extend many hundreds of kilometres and rise up to 3 km (1¾ miles) high. With little atmosphere to protect it, Mercury experiences extreme temperatures; in the Sun's full glare, the temperature can soar to a blistering 430°C (806°F) – hot enough to melt metal – but at night, it can plunge to a freezing –180°C (–292°F).

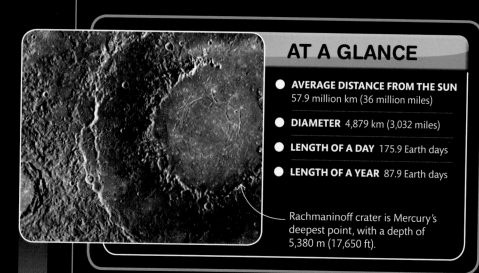

AT A GLANCE

- **AVERAGE DISTANCE FROM THE SUN**
  57.9 million km (36 million miles)

- **DIAMETER** 4,879 km (3,032 miles)

- **LENGTH OF A DAY** 175.9 Earth days

- **LENGTH OF A YEAR** 87.9 Earth days

Rachmaninoff crater is Mercury's deepest point, with a depth of 5,380 m (17,650 ft).

STATS AND FACTS

To the Romans, Mercury was the god of commerce, and the swift messenger of the gods. The planet Mercury shares his name because it is the fastest in the Solar System.

ORBITAL SPEED

km/h (in 1,000)	45	90	135	180

mph (in 1,000)		55		110

Mercury orbits the Sun at an average speed of 170,500 km/h (105,943 mph), making it the fastest orbiting planet.

BIGGEST CRATER

At 1,525 km (947½ miles) wide, the Caloris Basin is Mercury's largest crater.

km	1,000	1,500	2,000

miles	600		1,200

CRATERED WORLD

Mercury's moon-like surface is covered in craters, more than 390 of which are named after famous artists, writers, and composers. Beneath the barren rock, a large iron core makes up about three-quarters of the planet's interior.

THE CLOSEST PLANET TO THE SUN

SCORCHING WORLD
VENUS

Similar in size to Earth and made up of the same type of rocks and metals, Venus is often referred to as Earth's sister planet. It has vast canyons, rugged mountains, and towering volcanoes just like on Earth. But the similarities end there because Venus offers a far from friendly welcome! It has a dense, choking atmosphere composed mainly of toxic carbon dioxide, which traps so much heat that the surface of Venus is hot enough to melt lead. Thick clouds of harmful sulfuric acid swirl around the planet, carried by high speed winds.

AT A GLANCE

- **AVERAGE DISTANCE FROM THE SUN**
  108.2 million km (67¼ million miles)

- **DIAMETER** 12,104 km (7,521 miles)

- **LENGTH OF A DAY** 116.75 Earth days

- **LENGTH OF A YEAR** 224.7 Earth days

Maat Mons, the largest volcano on Venus, is 395 km (245 miles) wide and 8 km (5 miles) high.

STATS AND FACTS

Venus's dense atmosphere acts like a mirror, reflecting 80% of the radiation that reaches it from the Sun.

VOLCANOES

MORE THAN
1,600

DEADLY ATMOSPHERE

Carbon dioxide makes up 96.5% of Venus's atmosphere.

0   50   100

BRIGHTNESS

Venus is the second brightest object in the night sky after the Moon.

TEMPERATURE

Temperatures soar to 470°C (880°F), making Venus the hottest planet in the Solar System.

BARREN SURFACE

Using radar to penetrate Venus's thick atmosphere, this image reveals the rocky surface of the northern half of the planet. The bright spot in the centre is the planet's highest mountain range Maxwell Montes.

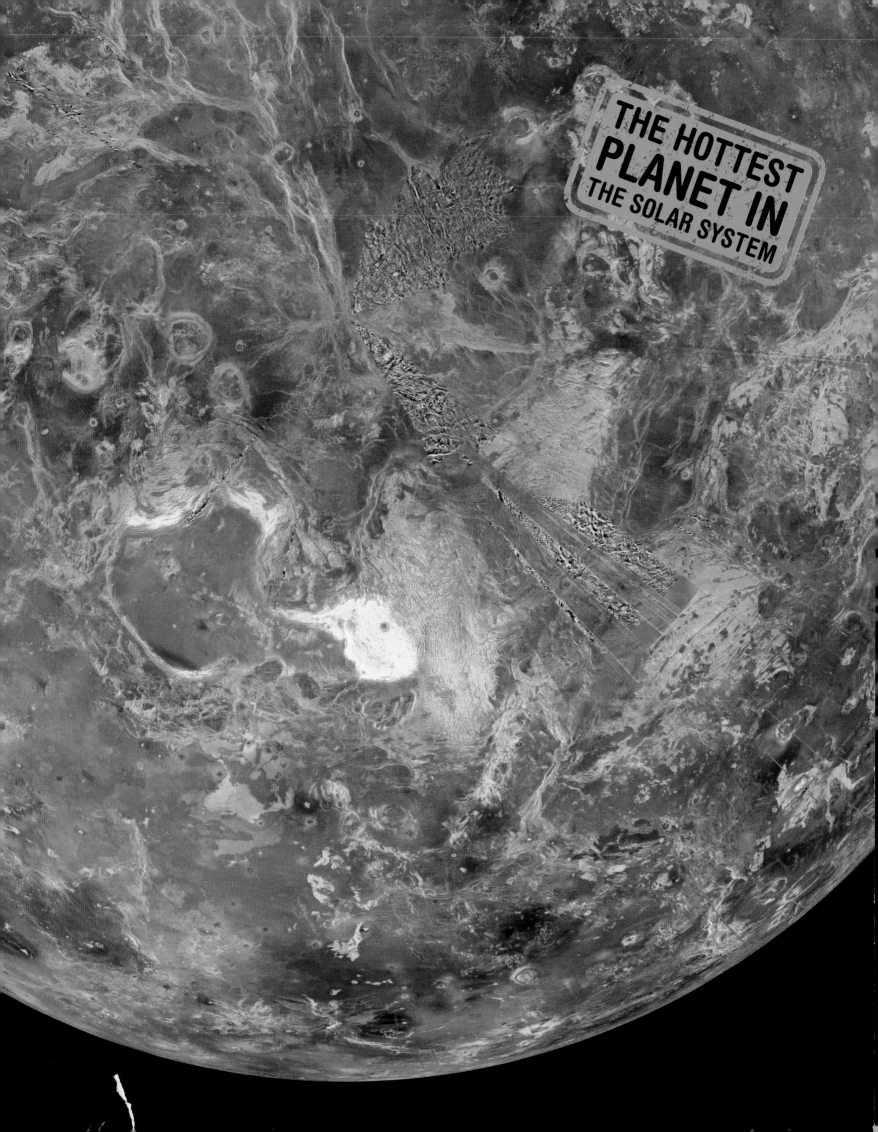

THE HOTTEST
PLANET IN
THE SOLAR SYSTEM

UNIQUE PLANET
EARTH

As far as we know, there is no planet like Earth – it is the only place where life exists. Our planet has all the right combination of ingredients needed to create and sustain life. Its distance from the Sun means it receives just the right amount of heat. Earth also has oceans of liquid water and a protective atmosphere. Our planet's surface is constantly changing as the forces of nature shape the land.

AT A GLANCE

- **AVERAGE DISTANCE FROM THE SUN**
  150 million km (93 million miles)

- **DIAMETER** 12,756 km (7,926 miles), at equator

- **AVERAGE SURFACE TEMPERATURE** 15°C (60°F)

- **LENGTH OF A YEAR** 365.25 days

STATS AND FACTS

Earth is the largest of the four rocky inner planets – bigger than Mercury, Venus, and Mars. About 29% of its surface is made up of land, but only 12% is inhabited by people.

SUNLIGHT

It takes about 8 minutes for light from the Sun to reach Earth's surface.

EARLY LIFE

Having existed for nearly 3.5 billion years, cyanobacteria are among Earth's oldest life forms.

EARTH'S CORE

At 5,600°C (10,100°F), Earth's inner core is as hot as the Sun's surface.

GREAT BARRIER REEF

The Great Barrier Reef, at more than 2,000 km (1,200 miles) long, is Earth's largest living structure.

AGE OF EARTH
4.54 BILLION YEARS

Earth's crust is divided into a number of major and minor interlocking tectonic plates.

Dynamic Earth

Earth is the only planet in the Solar System with a rocky crust that is broken into large pieces called plates. These plates shift slowly, but we usually don't notice their movements unless they produce an earthquake. Volcanoes, mountains, and deep ocean trenches form where the plate edges meet. The atmosphere that blankets our planet provides oxygen to breathe, traps heat, and shields us from the Sun's harmful rays.

Earth's atmosphere is made mostly of nitrogen and oxygen.

About 70% of Earth's surface is covered in liquid water.

THE ONLY PLANET KNOWN TO SUPPORT LIFE

THE LIVING PLANET

There is an astonishing variety of plants and animals almost everywhere on Earth, each adapted to survive in their varied habitats. Some environments, such as tropical rainforests and coral reefs, are teeming with life – but there are also living things in surprising places like deserts.

DESERT

TROPICAL RAINFOREST

CORAL REEF

POLAR ICE

EARTH FROM SPACE

At night, if you were looking at planet Earth from space, the signs of life would be immediately obvious. This spectacular view of Earth was captured by a weather satellite orbiting our planet. The bright areas are lights from the cities across Europe, north Africa, the Middle East, and Asia, home to some of the planet's 7.6 billion people.

EARTH'S COMPANION
THE MOON

The only world beyond Earth where astronauts have landed is the surface of the Moon. The Moon is an airless world with no liquid water and no atmosphere to shield it from harmful radiation or space debris. Unlike Earth's surface, which is continually changing due to weathering and movement of the crust, the Moon's surface barely changes, allowing thousands of craters to survive for billions of years.

The Moon's thin rocky crust is between 70 and 150 km (43–93 miles) thick.

AT A GLANCE

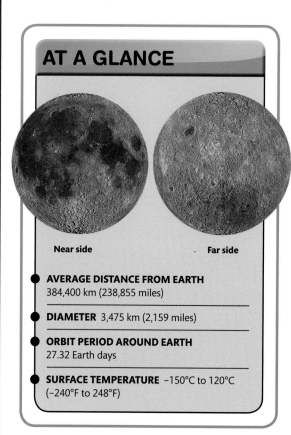

Near side   Far side

- **AVERAGE DISTANCE FROM EARTH**
  384,400 km (238,855 miles)

- **DIAMETER** 3,475 km (2,159 miles)

- **ORBIT PERIOD AROUND EARTH**
  27.32 Earth days

- **SURFACE TEMPERATURE** –150°C to 120°C
  (–240°F to 248°F)

Layered world

The Moon formed shortly after Earth, about 4.5 billion years ago. When it formed, the Moon was much hotter and its interior was molten. The heavy metals sank, forming an iron core, and lighter materials rose, creating the rocky outer layers. The dark areas on the lunar surface are known as maria, Latin for "seas". Early astronomers thought that these areas were covered with water. In fact, they are vast, ancient craters filled with solidified lava.

The entire surface is covered in a thick layer of fine dust that is several metres deep.

STATS AND FACTS

The Moon rotates once each time it orbits Earth, so we always see the same face – the near side. From Earth, many of its surface features are clearly visible. The Moon's far side, which can only be seen from space, is more heavily cratered and has fewer maria.

ORBITAL SPEED

The Moon orbits Earth at an average speed of about 3,680 km/h (2,290 mph).

MOONSHINE

Although the Moon is the brightest object in the night sky, it doesn't emit its own light but reflects the sunlight.

SIZE

The Moon is about a quarter of Earth's diameter, and is the fifth largest moon in the Solar System.

MOONQUAKES

Just as there are earthquakes on Earth, there are moonquakes, which can last for up to an hour.

There are thousands of craters on the Moon – the scars of asteroid and meteorite impacts.

CRATER FORMATION

Billions of years ago, the Moon was heavily bombarded by rocky material left over from the formation of the planets. Much of its surface is covered with craters of all sizes. The crater size and depth depends on the mass and speed of the impactor; small impacts leave bowl-shaped craters; larger impacts create massive craters with terraced walls and central mountains.

1

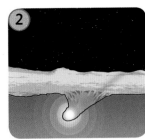

2

3

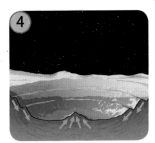

4

Space rock
Because there is no protective atmosphere, space rocks crash into the lunar surface.

Impact
The meteorite and some of the surrounding rock is vapourized by the energy released on impact.

Debris
Material from the impact site is thrown out, forming a layer of debris on the surrounding landscape.

Crater
In a large crater, the sides may slump, forming terraces, and the central floor may rebound, forming a peak.

The hot inner iron core is only about 480 km (300 miles) wide. It is kept solid by the pressure of the rock surrounding it.

The outer core is also made of iron, but may be molten. Scientists think it is about 100 km (60 miles) thick.

This inner mantle is made of rock but may be partly melted due to heat from the core.

The outer mantle makes up most of the Moon's interior. It consists of rocky materials like those on Earth.

MONTES APENNINUS

Located on the Moon's near side, Montes Apenninus is a spectacular example of the mountain ranges on the Moon. Stretching for about 600 km (373 miles), this mountain range contains about 3,000 peaks, including the Moon's tallest mountain, Mons Huygens, named after the Dutch astronomer Christiaan Huygens.

IMPACT CRATER

Named after a 16th century monk and astronomer, Giordano Bruno, this 22 km (13½ mile) wide impact crater is located on the Moon's far side. It is one of the youngest large lunar craters. Astronomers think it was formed within the past 10 million years – a mere blink of an eye compared to the 4.51 billion year lifetime of our Moon.

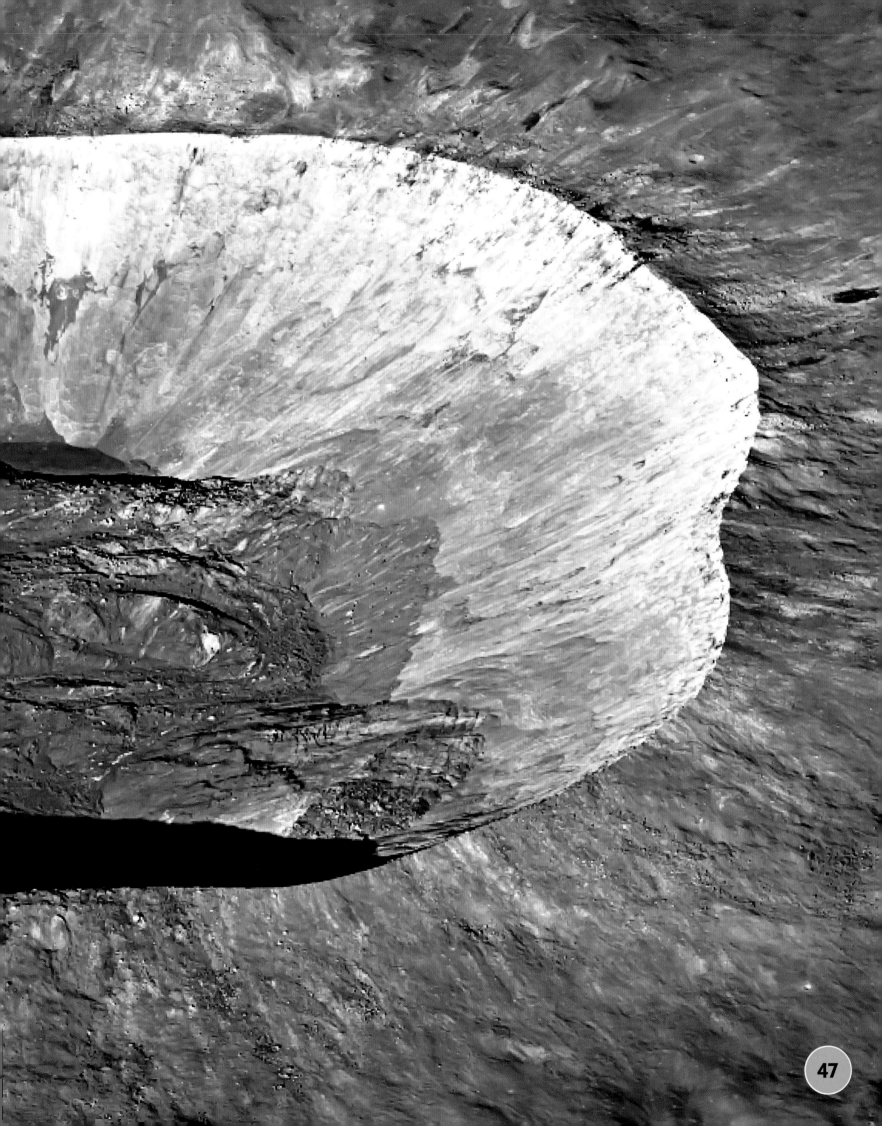

THE RED PLANET
MARS

Before about 3.8 billion years ago, Mars was warm and wet, but today the temperature has dropped and the river beds run dry. Since the first spacecraft passed in 1964, Mars has become the most probed planet in the Solar System. Evidence of ancient floodplains and crater lakes, together with the discovery of minerals that form only when water is present, lead scientists to think Mars once flowed with water and may have been home to extraterrestrial life.

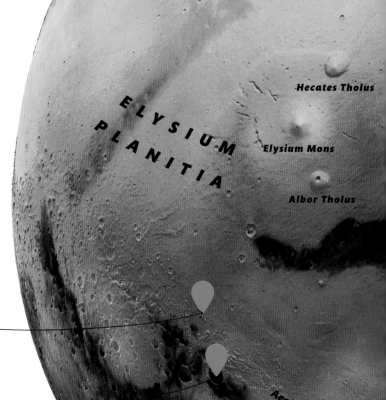

The Viking 2 lander touched down in a large plain called Utopia Planitia in 1976. It sent back the first images taken on the Martian surface.

VASTITAS

UTOPIA PLANITIA

Hecates Tholus

ELYSIUM PLANITIA

Elysium Mons

Albor Tholus

Herschel

Aeolis Mensae

TERRA CIMMERIA

AT A GLANCE

Hebes Chasma is a deep chasm located near a vast network of canyons on the Martian surface.

- **AVERAGE DISTANCE FROM THE SUN**
  228 million km (142 million miles)
- **DIAMETER** 6,792 km (4,220 miles)
- **LENGTH OF A DAY** 24.7 Earth hours
- **LENGTH OF A YEAR** 687 Earth days

Exploring Mars

In 1971, the lander from Mars 3 spacecraft became the first to achieve a successful landing on the planet's surface. Since then more than 40 spacecraft, landers, and rovers from six different nations have carried out investigations of the Red Planet. Some of the locations for the main landers and rovers visiting Mars are shown here.

NASA's InSight lander landed on a plain named Elysium Planitia in 2018 and was the first mission to use a robotic arm to grasp on-board instruments and place them on another planet.

The Curiosity rover landed in Gale Crater in 2012 and continues to explore the Martian surface.

STATS AND FACTS

The Solar System's biggest dust storms occur on Mars. It is known as the Red Planet because of its dusty red surface. The unique colour comes from iron minerals in the top layer of soil.

LARGE CRATER

Hellas Planitia on Mars is the Solar System's largest visible impact crater.

PHOBOS

Phobos orbits so closely that it travels around Mars in just 8 hours, shorter than one Martian day.

LAKE

The Mars Express orbiter has found an underground lake beneath Mars's south pole.

FUTURE MISSIONS

Missions are planned to take people to the Red Planet in the 2030s.

THE MOST PROBED PLANET IN THE SOLAR SYSTEM

The north polar ice cap is mainly frozen water with some frozen carbon dioxide, known as dry ice. Each summer the dry ice evaporates into the atmosphere.

The Phoenix lander arrived at a northern lowland plain named Vastitas Borealis in 2008. The first craft to land close to the north pole, Phoenix detected frozen water near the surface and observed snow falling from clouds in the planet's thin atmosphere, which is made up of mostly carbon dioxide.

PLANUM

BOREUM

BOREALIS

ARCADIA

PLANITIA

LYCUS SULCI

AMAZONIS

PLANITIA

Olympus Mons

Orcus
Patera

DAEDALIA

PLANUM

LUCUS

PLANUM

Gusev

Ma'adim Vallis

The Mars exploration rover Spirit landed at Gusev Crater in 2004 and travelled for nearly 8 km (5 miles) before losing power in 2010.

SURFACE FEATURES

For decades spacecraft have mapped the surface of Mars, including polar ice caps, volcanoes, valleys, canyons, and cliffs. Key features have been photographed to build a clear picture of this varied planetary landscape.

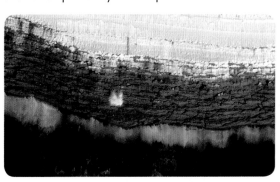

ICY AVALANCHE
The Mars Reconnaissance Orbiter captured the moment an avalanche occurred at the northern polar ice cap. The avalanche created clouds of dust on impact with the ground.

MIGHTY MOUNTAIN
The tallest known mountain in the Solar System is this shield volcano on Mars. At 25 km (16 miles) high, Olympus Mons is almost three times the height of Mount Everest.

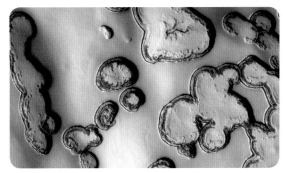

SOUTHERN PITS
During summer in Mars's southern hemisphere, the layer of frozen carbon dioxide on the ground starts to evaporate, revealing the edges of the flat-floored pits.

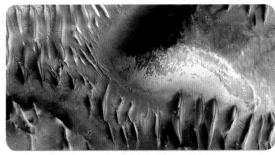

SAND DUNES
Part of the Martian surface is covered by sand dunes, as photographed by the Mars Reconnaissance Orbiter. The shapes of the dunes indicate the directions of the winds that sculpted them.

SPACE CANYON

Named after the Mariner 9 spacecraft that discovered it in 1971, Valles Marineris is a vast, meandering canyon system that slices across the face of Mars, making it the longest canyon in the Solar System. It is more than 4,000 km (2,485 miles) long – as long as the distance from New York to Los Angeles in the USA. This image, taken by NASA's Viking 1, shows just part of the canyon.

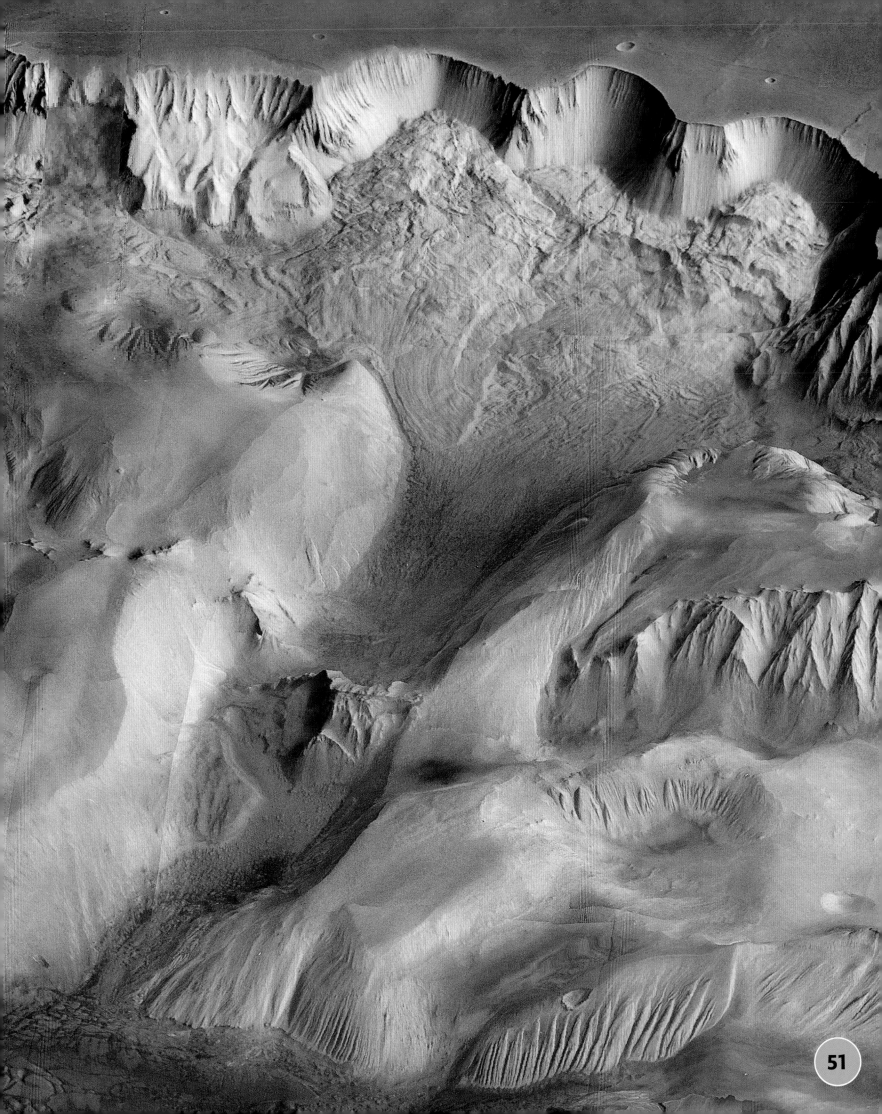

SHOOTING STARS
THE PERSEIDS

MOST VISIBLE METEOR SHOWER FROM EARTH

The Perseid meteor shower is a spectacular event caused by Earth's orbit taking it through a cloud of debris left behind by a comet. The tiny particles of rock, dust, and ice hit Earth's atmosphere at high speed, which makes them burn up as meteors and appear as fiery streaks of light across the night sky. Meteors are often called shooting stars, although they are not actually stars. The Perseid meteor shower is known for its fast-moving, bright meteors each year.

AT A GLANCE

- **PEAK ACTIVITY** Mid-August
- **VIEWABLE** Northern Hemisphere, best after midnight
- **SOURCE COMET** 109P/Swift-Tuttle
- **FIRST RECORDED OBSERVATION** 36 CE in ancient China

STATS AND FACTS

Meteor showers appear to come from the same point in the night sky, called the radiant. The constellation in which the radiant is located gives the meteor shower its name; in this case the constellation is Perseus.

SPEED

The Perseids travel through the atmosphere at a speed of about 212,400 km/h (133,200 mph).

km/h (in 1,000)	200	400
mph (in 1,000)	125	250

RECORD BREAKER

A shower in 1993 had the most number of Perseid meteors ever spotted – over 300 per hour.

PEAK NUMBER

100

The maximum number of meteors normally seen per hour is about 100.

FIREBALL

Although most meteors are only the size of a grain of sand, some can be the size of a marble. They look like mini fireballs in the night sky like this Perseid meteor.

METALLIC ROCK
HOBA

Earth is often bombarded by pieces of space debris that are too big to burn up in its atmosphere. Known as meteorites, many fall into the ocean or are too small to detect on land. A few disintegrate upon hitting the ground, leaving evidence in the form of an impact crater. But this was not the case with the largest and heaviest of the meteorites discovered so far – Hoba. Some scientists believe that Hoba's large, flat sides may have acted like a brake as it fell through Earth's atmosphere, slowing it down enough to land in one piece. It still lies where it fell in southern Africa about 80,000 years ago, too heavy to be moved.

AT A GLANCE

- **LOCATION** Namibia, southern Africa
- **DISCOVERED** In 1920 on Hoba West farm by the owner of the land
- **AGE** About 4,300 million years old
- **SIZE** 2.7 m × 2.7 m × 0.9 m (9 ft × 9 ft × 3 ft)

STATS AND FACTS

More than 40,000 meteorites have been discovered on Earth so far. Most are pieces of asteroids that were broken apart during a collision in space. Many meteorites are lumps of stone, while others like Hoba consist of iron and nickel. The rarest meteorites are a mixture of stone and iron.

MIGHTY METAL

Hoba weighs as much as 10 African elephants.

COMPOSITION

The meteorite consists of 84% iron and 16% nickel, with small traces of cobalt.

WEIGHT

At 60 tonnes, Hoba is the heaviest meteorite on Earth.

tonnes	30	60	90

At 37 tonnes, El Chaco is the second heaviest meteorite on Earth.

THE WORLD'S BIGGEST INTACT METEORITE

RUSTING AWAY

This close-up view of the Hoba meteorite shows its ridged exterior and streaks of orange-red where rust has eaten away at the iron it contains. Since its discovery, the meteorite has lost an estimated 6 tonnes to rust and souvenir hunters.

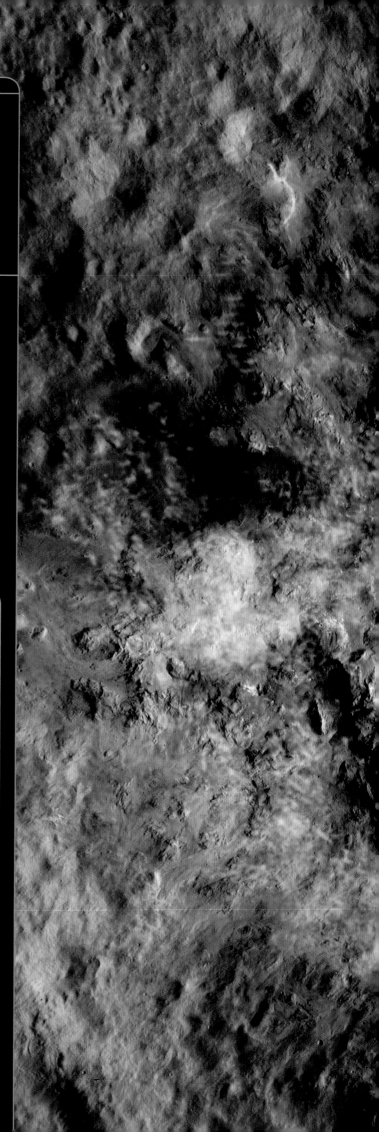

ROCKY DWARF PLANET
CERES

The Asteroid Belt is a vast region of rocky debris between the orbits of Mars and Jupiter. The asteroids are rocks left over from the formation of the Solar System 4.6 billion years ago; the most massive, and the first to be discovered, was Ceres. Because Ceres has a spherical shape, it is now classified as a dwarf planet. In 2015, NASA's Dawn spacecraft became the first mission to visit Ceres, revealing a heavily cratered surface, a volcano-like mountain, and many startling white patches on the surface, which might be a kind of salt from an underground ocean.

AT A GLANCE

- **AVERAGE DISTANCE FROM THE SUN**
  414 million km (257 million miles)

- **DIAMETER** 952.4 km (592 miles),
  at equator

- **LENGTH OF A DAY** 9 Earth hours

- **LENGTH OF A YEAR** 1,682 Earth days

  This prominent impact crater
  on Ceres is called Occator.

STATS AND FACTS

While there are more than 780,000 objects in the Asteroid Belt, Ceres is the largest and contains one quarter of all the material in the Asteroid Belt. The other objects range in size from 530 km (329 miles) to just 10 m (33 ft) in diameter.

DISCOVERY

Italian priest Giuseppe Piazzi discovered Ceres in 1801 while looking for a star.

SURFACE AREA

Ceres's surface area is slightly less than the area occupied by India on Earth.

WATER

Ceres may have once had water in the past, and some may still exist.

MOUNTAIN

Ahuna Mons, the largest mountain on Ceres, is about 5 km (3 miles) high.

BRIGHT SPOTS

The Haulani Crater on Ceres is about 34 km (21 miles) wide. It contains one of several bright spots discovered on this dwarf planet, as seen in this image taken by NASA's Dawn spacecraft. The bluish colour denotes features on Ceres that are relatively young.

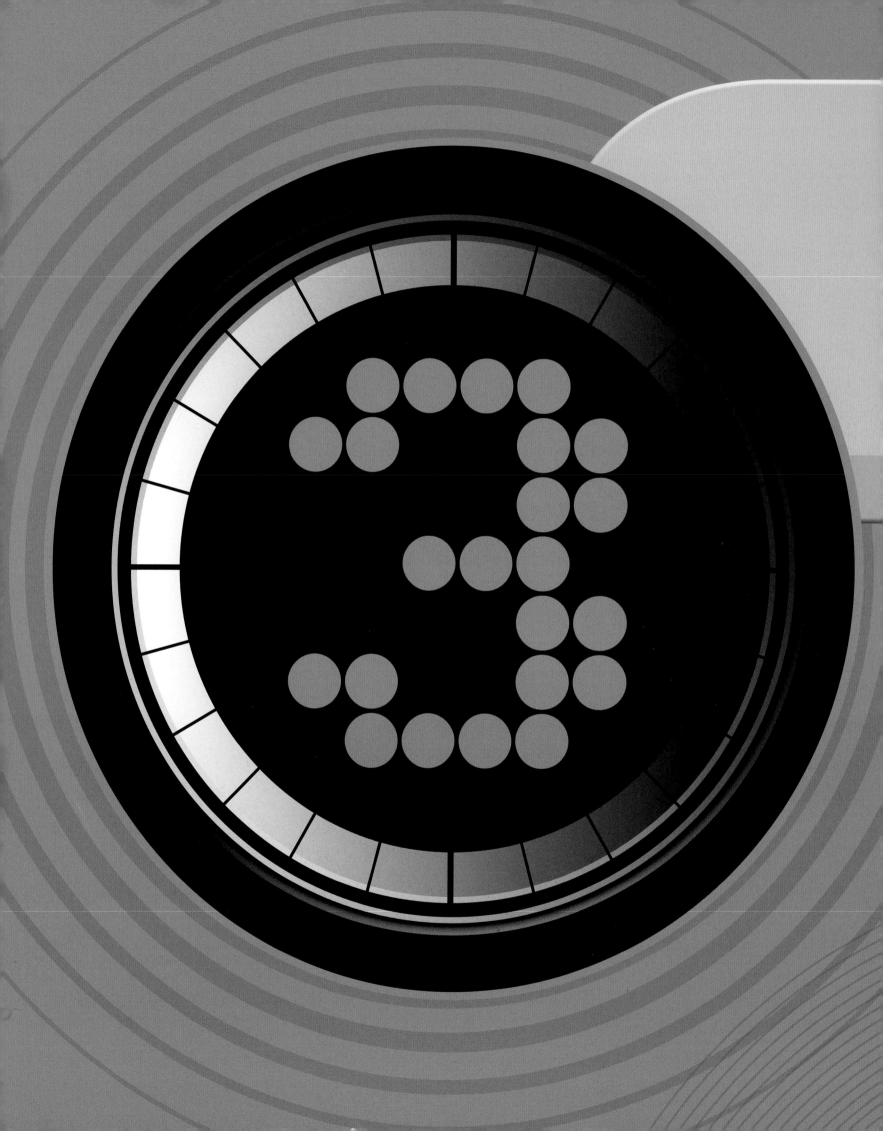

OUTER SOLAR SYSTEM

Beyond the Asteroid
Belt far from the Sun is
deep, dark space. This remote
region is freezing cold and
dominated by the four giant
planets: Jupiter, Saturn, Uranus,
and Neptune. Each one
features its own family
of moons and rings of dust
and icy chunks.

GIANT PLANET
JUPITER

Jupiter is more than twice as massive as all the other planets put together. It is a gas giant made up of mostly hydrogen and helium, with colourful bands and swirling storms visible on its surface. Jupiter's extreme size and brilliant glow combine to make it the third brightest object in the night sky after the Moon and Venus.

THE LARGEST PLANET IN OUR SOLAR SYSTEM

AT A GLANCE

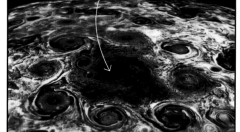

Storms above Jupiter's north pole are seen here in infrared light.

- **AVERAGE DISTANCE FROM THE SUN**
  778.6 million km (484 million miles)

- **DIAMETER** 142,984 km (88,846 miles), at equator

- **LENGTH OF A DAY** 9.9 Earth hours

- **LENGTH OF A YEAR** 11.9 Earth years

STATS AND FACTS

Italian astronomer Galileo Galilei discovered four of Jupiter's moons in 1610. Since 1973, nine spacecraft have visited the planet, discovering a family of more than 70 moons. The largest moon is Ganymede, which is bigger than Mercury.

WIDTH

Jupiter is nearly 11 times wider than Earth at its equator.

SHORT DAY

Jupiter has the shortest day in the Solar System.

WIND SPEEDS

Winds can reach speeds of more than 400 km/h (250 mph) along Jupiter's equatorial region.

MAGNETIC FIELD

The magnetic field on Jupiter is at least 20 times more powerful than that on Earth.

The atmosphere's outer cloud layer is just 50 km (31 miles) thick.

Cloud cover

Jupiter's colourful cloud bands are the result of different chemicals, reflecting the Sun's light. Winds blowing in both directions create the striking patterns and cloud bands that wrap around the planet. Scientists think that a vast ocean of liquid metal may lie beneath its thick cloud cover.

NORTHERN HEMISPHERE

The swirling, turbulent clouds in Jupiter's northern hemisphere can be seen in this colour-enhanced image taken by NASA's Juno spacecraft. The patterns are caused by rising and falling areas of gas in the atmosphere. The higher clouds in white are seen casting shadows on the clouds below them. The white ovals are rotating storms.

The darker cloud bands, called belts, are areas of sinking gas.

Many of the bands are wider than Earth.

Four broad but faint rings made up of small pieces of dust encircle the planet.

The lighter cloud bands are called zones. They are areas of rising gas.

"1,300 Earths could fit into Jupiter with room to spare."

Jupiter's most striking feature is the Great Red Spot – a massive anticlockwise-rotating storm.

AURORAS ON JUPITER

Spectacular light displays, called auroras, are sometimes seen in the atmosphere surrounding Jupiter's poles. Their visible glow is the result of high-energy particles from space colliding with gas atoms. Jupiter's auroras have more than 100 times the intensity of those on Earth. This image taken by the Hubble Space Telescope in 2007 shows Jupiter's auroras in blue.

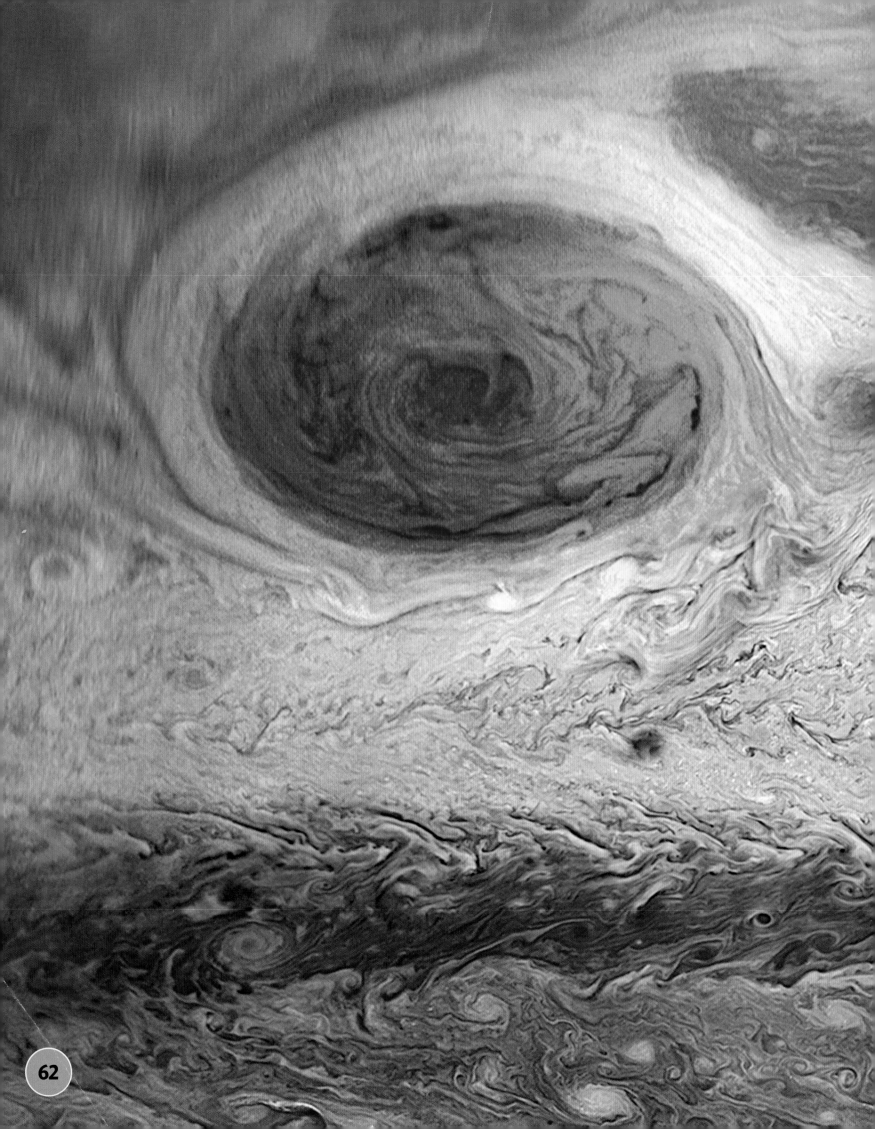

MEGA STORM

Jupiter's Great Red Spot is the Solar System's longest-raging storm. People have observed it for more than 200 years. It appears to have got smaller since it was first examined closely by Voyager 1 in 1979, but at 16,350 km (10,159 miles) wide, the Great Red Spot is still big enough to swallow Earth. Winds swirl around its centre at speeds of up to 680 km/h (422 mph).

ICY MOON
EUROPA

The fourth largest moon of Jupiter, Europa is an intriguing world. The moon's outer crust is made of frozen water and, strangely, is almost completely free of craters. Cracks in the surface create mysterious double ridges and cause ice quakes. Scientists believe that beneath Europa's icy crust lies a single ocean that flows across the entire moon. This hidden ocean may be home to alien life.

Icy depths
The Galileo spacecraft orbited Jupiter and its moons between 1995 and 2003. Data from the mission suggests there is a vast ocean of liquid water beneath Europa's thick icy crust. Scientists think this sub-surface ocean could contain more than twice as much water as all Earth's oceans.

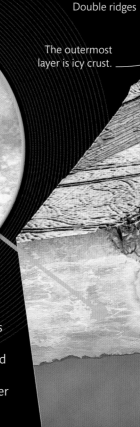
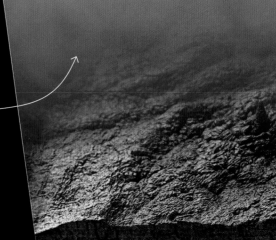

Double ridges

The outermost layer is icy crust.

AT A GLANCE

This photograph, taken by the Galileo spacecraft, shows long fractures in Europa's thin ice crust.

DISTANCE FROM JUPITER
671,000 km (417,000 miles)

DIAMETER About 3,122 km (1,940 miles)

ORBITAL PERIOD 3.5 Earth days

SURFACE TEMPERATURE
Below –160°C (–256°F) at the equator and below –220°C (–364°F) at the poles

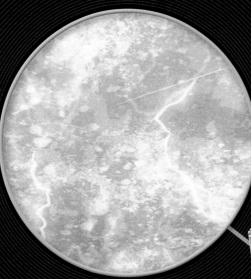

ICE FRACTURES
Europa's icy shell contains many ridges and cracks where large blocks of ice have moved and sometimes overturned before refreezing into new positions. This suggests there may be flowing water beneath the surface.

If there is an ocean beneath the surface, it is likely to be warmed by tidal forces affecting both the water and Europa's core.

STATS AND FACTS

ABOUT
25 KM
(15½ MILES)
THICKNESS OF ICE SHELL

To find out more about Europa both NASA and ESA are planning missions to explore the moon's icy interior.

DEPTH

Europa's ocean may be up to 150 km (93 miles) deep.

km	100	200
miles	62	124

The deepest point in Earth's Pacific Ocean lies 11 km (6¾ miles) beneath the water surface.

PLUMES

Scientists suspect that huge plumes of water may erupt from Europa's surface.

LARGE MOON

Europa is the sixth largest moon in the Solar System.

GALILEAN MOONS

Jupiter's four largest moons are named after the Italian astronomer Galileo Galilei, who spotted them in 1610 while observing Jupiter. They were the first objects to be discovered orbiting a planet other than Earth, and have very different characteristics from each other. The Galilean moons are shown here in order of increasing distance from the planet.

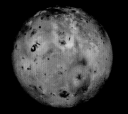

Io

More than 400 active volcanoes make Io the Solar System's most volcanic object. Some of them spurt fountains of lava more than 100 km (62 miles) above its surface.

Europa

Jupiter's ice-covered moon Europa is a little smaller than Earth's Moon. It has the youngest surface, with an age of between 40 and 90 million years.

Ganymede

The largest moon in the Solar System, Ganymede is bigger than the planet Mercury. It is composed of rock and ice, with shallow craters and long ditches etched into its surface.

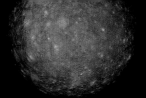

Callisto

The second largest moon of Jupiter, Callisto has a surface scarred by ancient encounters with meteorites. Some parts of its cratered surface are about 4 billion years old.

Scientists think that plumes of water and water vapour may erupt through the ice, rising many kilometres above the surface.

The smooth surface is criss-crossed by dark red lines, which may be the result of minerals seeping through from below and freezing.

In some parts of the surface, the ice crust has broken up.

HYDROTHERMAL VENT

On Earth, openings on the oce floor expel swirling streams of g minerals, and hot water, heated the core below. Scientists susp that vents like these may exist Europa, warming the ocean a enriching it with the mineral necessary for life to exist there

RINGED PLANET
SATURN

Circled by its spectacular icy rings and more than 60 known moons, Saturn is one of the wonders of the Solar System. Like Jupiter, it is a gas giant, consisting mostly of the lightest chemical elements – hydrogen and helium. It is also the least dense of all the planets, making it the only one that is lighter than water, and the least spherical planet – it spins so rapidly that it bulges at the equator. While Saturn normally looks tranquil, huge white storm clouds occasionally erupt in its atmosphere. Powerful winds whirl around its north and south poles, like raging hurricanes. Over the north pole, streaming winds form a mysterious hexagonal pattern.

Icy objects in Saturn's rings vary from tiny grains to house-sized boulders.

AT A GLANCE

- **AVERAGE DISTANCE FROM THE SUN**
  1,433.5 million km (890⅗ million miles)

- **DIAMETER** 120,536 km
  (74,898 miles), at equator

- **LENGTH OF A DAY** 10.7 Earth hours

- **LENGTH OF A YEAR** 29.46 Earth years

STATS AND FACTS

Saturn has been known to astronomers since ancient times but the rings were first observed in 1610 by Italian astronomer Galileo Galilei. In modern times, four spacecraft have visited Saturn, sending back incredible details of the planet, its moons, and rings.

NORTH POLE HEXAGON

km (in 1,000)	10	20
miles (in 1,000)	6	12

At about 13,800 km (8,575 mile) long, each side of Saturn's hexagonal cloud pattern is wider than Earth.

RINGS

Compared to the size of the planet, Saturn's vast ring system appears paper-thin.

LIGHTNING

Lightning on Saturn is 10,000 times more powerful than on Earth.

ICY HALO
Saturn's colossal
rings are made of
billions of pieces of dirty ice
that lie in an almost flat plane
around the planet. The ice
reflects the light from the Sun,
which is why you can see it; the
dark areas are gaps in the rings.

PLANET-LIKE MOON

Big lakes of liquid ethane and methane lie beneath the opaque orange atmosphere of Titan, Saturn's largest moon. Chemicals like ethane and methane are found as natural gases on Earth, but on Titan's cold surface, they exist as liquids, forming lakes and rivers as shown in this artist's illustration. Titan has surface features reminiscent of the young Earth and, like Earth, nitrogen-rich air and a weather system.

SYSTEM OF RINGS

A set of 13 faint rings encircles this ice giant. The nine inner rings are narrow and dark grey, while the outer rings often appear reddish and blue. Unlike Saturn's rings, which are mostly ice, Uranus's ring system contains dust and dark rocky material.

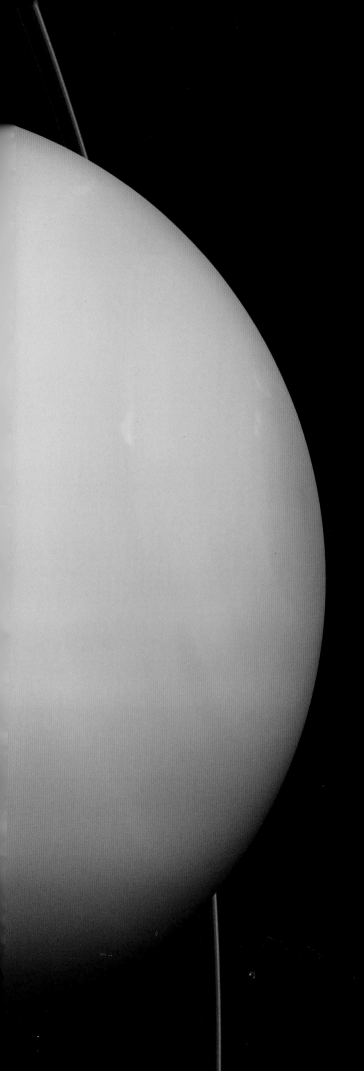

SIDEWAYS PLANET
URANUS

Uranus is a unique planet – it orbits the Sun tilted on its side. As a result, its polar regions experience 42 years of darkness followed by 42 years of continuous sunlight. The planet appears featureless apart from indistinct cloud bands of methane and sulfurous gases. Under the atmosphere, the gas merges into a vast ocean of water, methane, and ammonia, with a small rocky core at the centre. Methane gas in its upper atmosphere absorbs the red part of the sunlight, giving the planet its distinctive blue colour. We know of at least 27 moons orbiting Uranus.

AT A GLANCE

AVERAGE DISTANCE FROM THE SUN
2.87 billion km (1¾ billion miles)

DIAMETER 51,118 km (31,763 miles), at equator – about 400% that of Earth

LENGTH OF A DAY 17.2 Earth hours

LENGTH OF A YEAR 83.8 Earth years

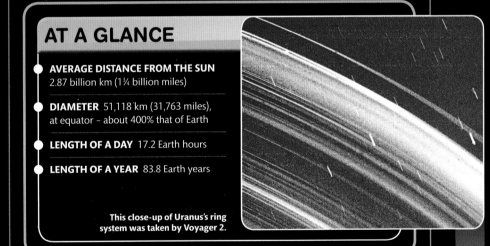

This close-up of Uranus's ring system was taken by Voyager 2.

STATS AND FACTS

Because Uranus has such an extreme sideways tilt, it is one of only two planets that spin backwards; the other is Venus. Its unusual sideways tilt also causes it to have the most extreme seasons of any planet in the Solar System.

AXIS TILT

Uranus's axis is tilted at 97.8°.

0°   50°   100°

Earth's axis is tilted at 23.4°.

TALLEST CLIFF

At 20 km (12 miles) high, Verona Rupes on Uranus's moon, Miranda, is the Solar System's tallest cliff.

DIAMOND LAYER

Scientists think that Uranus may have a layer of diamond deep inside.

THE WINDIEST PLANET IN OUR SOLAR SYSTEM

BLUE PLANET

Neptune owes its vivid blue
colour to traces of methane
gas in its upper atmosphere –
methane absorbs long light
wavelengths such as red and
orange while reflecting short
wavelengths such as blue.

WINDY WORLD
NEPTUNE

On ice-cold Neptune, winds whip through the atmosphere at supersonic speeds – far faster than ever encountered on Earth – making Neptune the Solar System's windiest planet. It is also the Solar System's most distant planet – about 30 times farther away from the Sun than Earth. The ice giant is mainly a mixture of water, methane, and ammonia, with rocky material concentrated towards the middle. Its atmosphere is made up mostly of hydrogen and helium. Since its discovery in 1846, Neptune has made only one complete orbit around the Sun.

AT A GLANCE

- **AVERAGE DISTANCE FROM THE SUN**
  4.5 billion km (2¾ billion miles)

- **DIAMETER** 49,528 km (30,775 miles), at equator – about 388% that of Earth

- **LENGTH OF A DAY** 16.1 Earth hours

- **LENGTH OF A YEAR** 163.7 Earth years

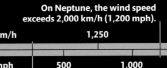

These are high altitude clouds in Neptune's atmosphere.

STATS AND FACTS

In 1989, the Voyager 2 spacecraft discovered a storm in Neptune's atmosphere that was the size of Earth. The craft also confirmed that the planet has a small system of five dark rings surrounding its equator.

WIND SPEED

On Neptune, the wind speed exceeds 2,000 km/h (1,200 mph).

km/h	1,250	2,500
mph	500	1,000	1,500

On Earth, the highest recorded wind speed is 408 km/h (253 mph).

GEYSERS

Neptune's moon Triton has active geysers erupting plumes of gas and dust up to 8 km (5 miles) high.

km	5	10
miles	3	6

NEPTUNE'S MOON

In 1989, NASA's Voyager 2 discovered huge geysers on Triton, the largest of Neptune's 14 known moons. The temperature and pressure changes beneath Triton's surface lead to jets of gas erupting from its surface, as shown in this artist's impression.

SPACE SNOWBALL
COMET McNAUGHT

In the vast, freezing outer reaches of the Solar System trillions of icy comets orbit the Sun. They are too distant to be seen, but every so often one travels closer to the Sun, triggering a dramatic change. As it warms up, it releases huge amounts of gas and dust, forming a vast cloud that is visible from Earth. This was the case with Comet McNaught. In early 2007, it became the brightest comet to fly past our planet in 40 years. The length of its tail was estimated to be more than one and a half times the distance between Earth and the Sun. It's not likely to return to the inner planets for thousands of years.

AT A GLANCE

SIZE OF NUCLEUS (CORE)
25 km (15 miles) across

AT BRIGHTEST 13–14 January 2007

TIME TAKEN TO ORBIT THE SUN
92,600 Earth years

CLOSEST APPROACH TO THE SUN
About 25.5 million km (15⅘ million miles)

Colourized image highlights the details in the head of the comet.

STATS AND FACTS

Comet McNaught was first discovered in August 2006 by British-Australian astronomer Robert McNaught as he searched for near-Earth asteroids. The comet became visible with the unaided eye in January 2007, when it reached its closest point to the Sun.

GAS TAIL

Comet McNaught had a gas tail longer than 260 million km (161½ million miles).

km (in million)	300	600
miles (in million)	200	400

Comet Hyakutake had the longest measured comet gas tail at 570 million km (355 million miles) long.

FORMATION

Comets formed 4.6 billion years ago from the leftover material of the Solar System.

NUCLEUS

The main body of the comet, the nucleus, is made of ice, rock dust, and frozen gases.

DEEP FREEZE
PLUTO

Located in the Kuiper Belt, Pluto is 40 times farther away from the Sun than Earth. It was discovered in 1930, and for 76 years was considered the Solar System's ninth planet until astronomers reclassified it to dwarf planet status. In 2015, New Horizons – the first spacecraft to reach Pluto – revealed a far-from-featureless world. The amazing close-up images of Pluto showed sprawling icy plains, large cliffs, and towering mountains. New Horizons also photographed Pluto's five moons. The largest, called Charon, lies very close to Pluto; scientists think it may have once been part of Pluto, broken off by a collision.

AT A GLANCE

- **AVERAGE DISTANCE FROM THE SUN**
  5.9 billion km (3⅗ billion miles)

- **AVERAGE SURFACE TEMPERATURE**
  –240°C (–400°F)

- **LENGTH OF A DAY** 153 Earth hours

- **LENGTH OF A YEAR** 248 Earth years

This vast, crater-free region on Pluto is called Sputnik Planitia. It is the largest glacier in the Solar System.

STATS AND FACTS

Pluto was discovered in 1930 by American astronomer Clyde Tombaugh. In his search for the existence of a ninth planet, he photographed the same patches of sky on different nights until he found a tiny dot of light that moved.

DIAMETER

Smaller than Earth's Moon, Pluto is 2,377 km (1,477 miles) in diameter.

km	2,000	4,000
miles	1,200	2,400

Earth's Moon has a diameter of 3,475 km (2,159 miles).

ORBIT

Pluto's elongated orbit is tilted by 17° – more than the orbit of any major planet.

CHARON

Pluto's moon Charon is half the size of Pluto. No planet has a moon so relatively big.

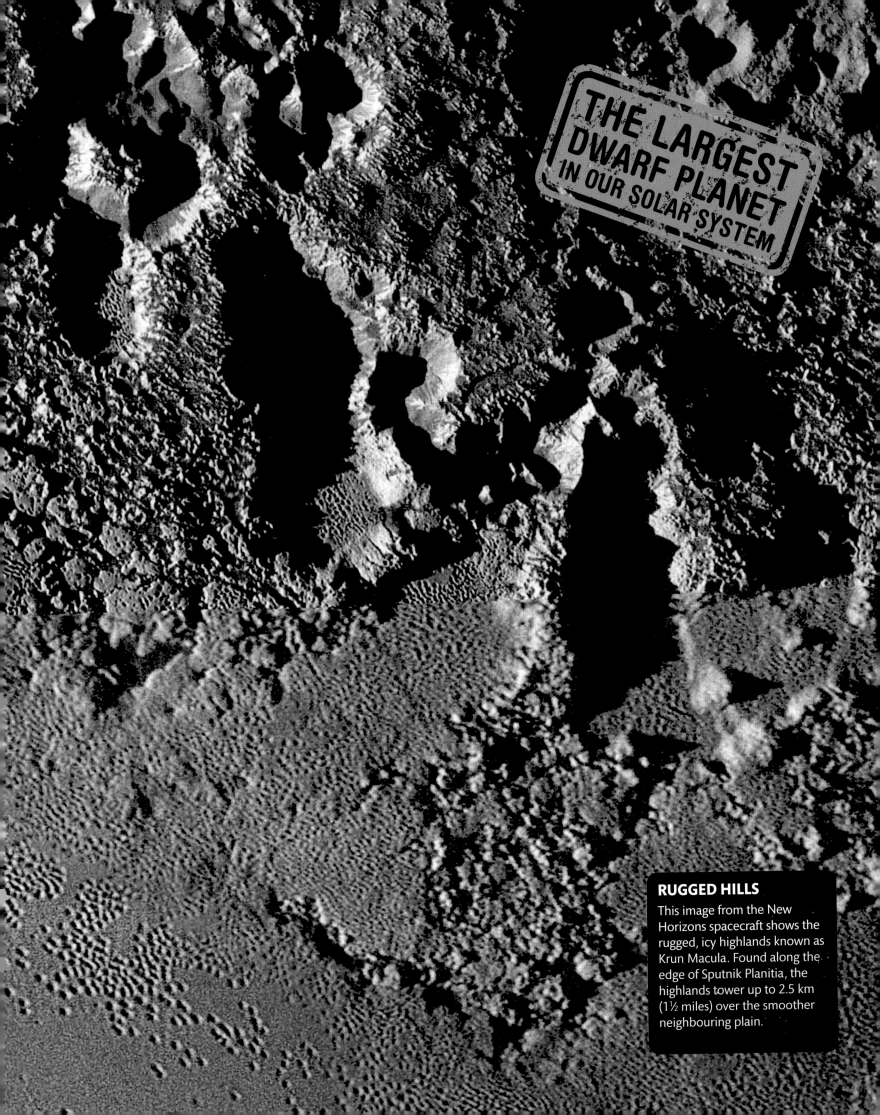

THE LARGEST DWARF PLANET IN OUR SOLAR SYSTEM

RUGGED HILLS

This image from the New Horizons spacecraft shows the rugged, icy highlands known as Krun Macula. Found along the edge of Sputnik Planitia, the highlands tower up to 2.5 km (1½ miles) over the smoother neighbouring plain.

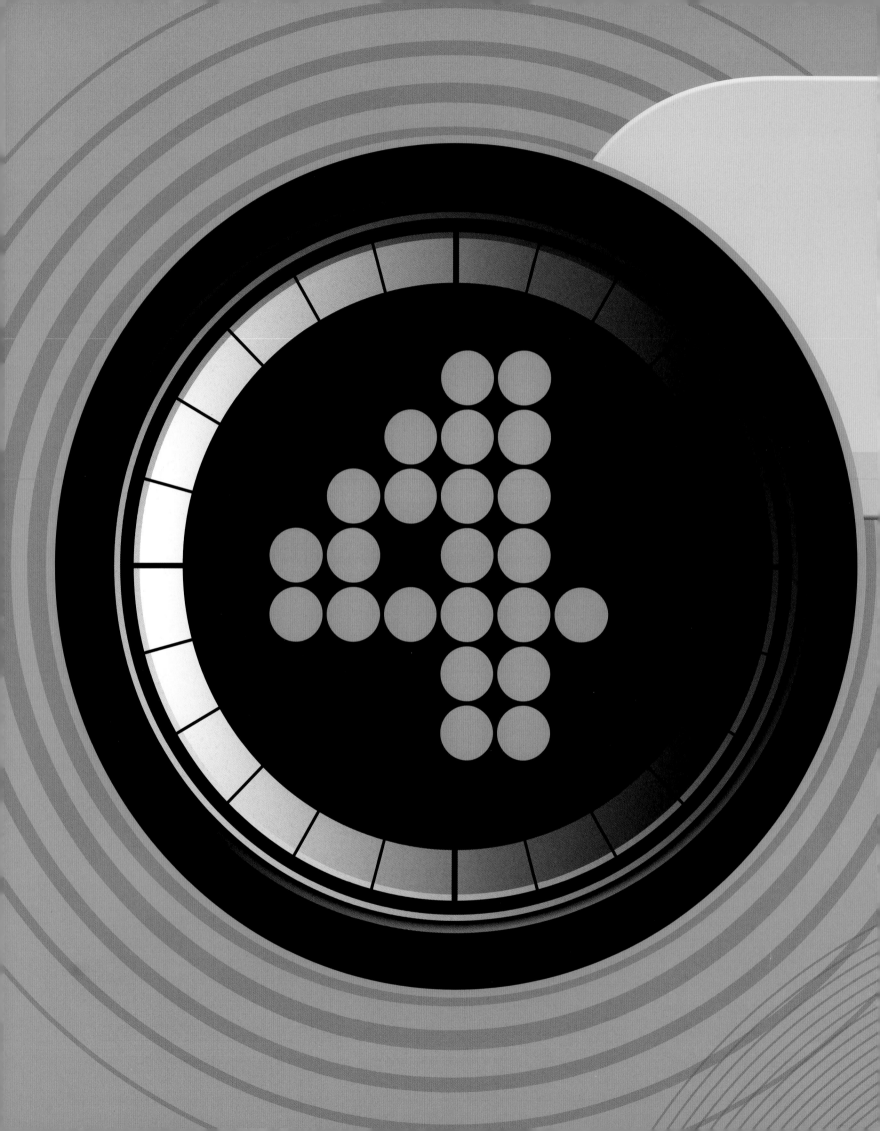

STARS AND EXOPLANETS

Our night sky is illuminated by thousands of sparkling stars. But these are just a tiny proportion of the trillions found throughout the Universe. Like our Sun, many of the stars we see have their own orbiting worlds, called exoplanets.

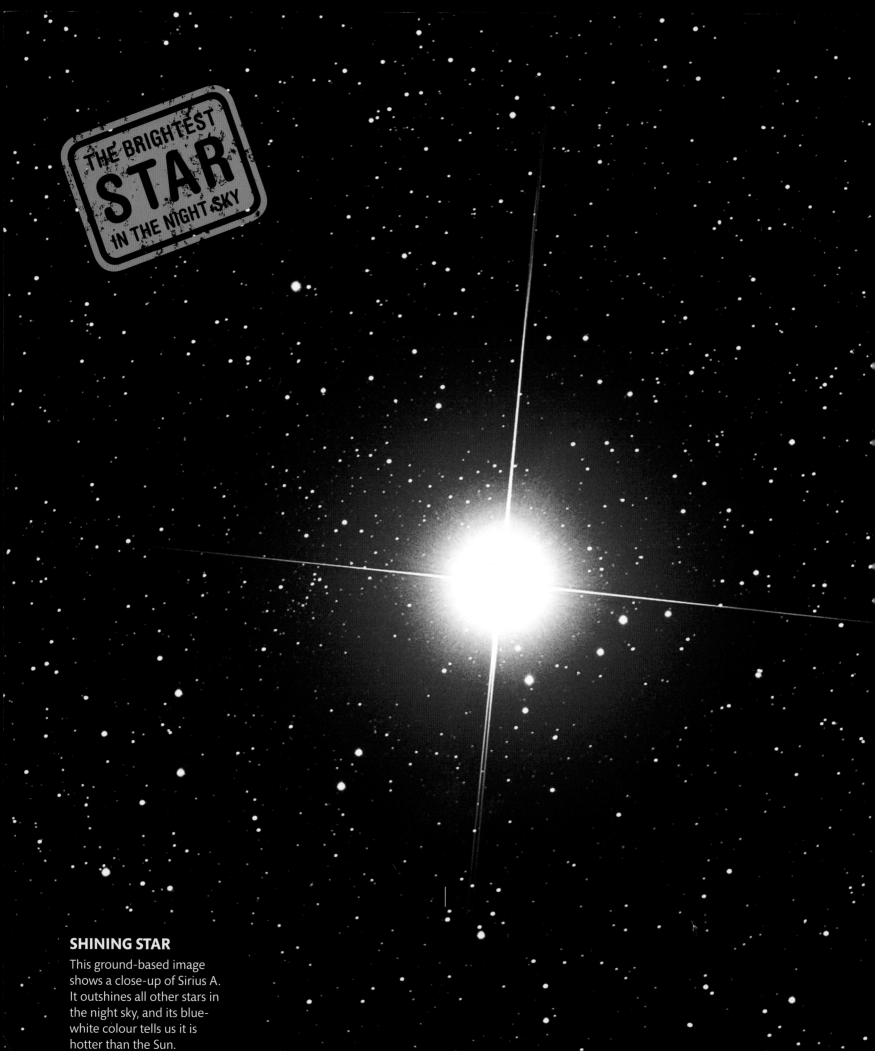

SHINING STAR

This ground-based image shows a close-up of Sirius A. It outshines all other stars in the night sky, and its blue-white colour tells us it is hotter than the Sun.

THE DOG STAR

SIRIUS A

Most of the stars we see in the Milky Way have one or more companion stars. Often one star is heavier than the other. Sirius A is not only heavier than its partner but it is twice as massive as the Sun, and because of its luminosity and closeness to Earth, it is by far the brightest star in the night sky. Its tiny companion Sirius B was the first white dwarf star to be discovered – the collapsed remains of a giant star that is 10,000 times fainter than its bigger partner. About 120 million years ago, Sirius B would have outshone its companion star. Bound by gravity, the two stars take about 50 years to orbit each other.

AT A GLANCE

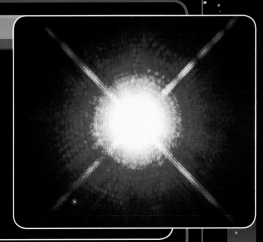

- **STAR TYPE** Main sequence white star
- **DISTANCE FROM EARTH** 8.6 light years
- **AGE** 200–300 million years old
- **SURFACE TEMPERATURE** 10,000°C (18,032°F)

Sirius A with its fainter companion Sirius B

STATS AND FACTS

Also known as the Dog Star because of its location in the constellation Canis Major (Latin for the Big Dog), Sirius has been observed by people all over the world for centuries. Its fainter companion star is often called the Pup.

LUMINOSITY

Sirius A is 25 times more luminous than the Sun.

DIAMETER

With a diameter of 2.4 million km (1½ million miles), Sirius A is about twice the size of the Sun.

COMPANION STAR

Sirius A's companion star Sirius B was discovered in 1862.

FLAG OF BRAZIL

Sirius is one of the 27 stars on Brazil's flag. It depicts the state of Mato Grosso.

SUPERSIZED STAR
VY CANIS MAJORIS

All stars shine, but VY Canis Majoris stands out from most. With room to fit 3 billion Suns inside, this star is one of the largest known in the Milky Way galaxy. Scientists were able to confirm its distance and huge size in 2008 and 2012 by observing the microwaves it emits. About 1,400 times wider than the Sun, it would engulf the planets out as far as Jupiter if placed in the centre of our Solar System. This red hypergiant (a subclass of red supergiants) is expected to explode violently within the next 100,000 years.

Sun   Canis Majoris

AT A GLANCE

- **STAR TYPE** Red hypergiant

- **DISTANCE FROM EARTH**
  About 3,900 light years

- **MAGNITUDE** Varies between
  6.5 and 9.6

- **FIRST RECORDED** In 1801 by French
  astronomer Jérôme Lalande

Comparison of our Sun and
VY Canis Majoris

STATS AND FACTS

VY Canis Majoris is a single star, which makes its mass impossible to measure. Stars of a similar type and age are usually 15–20 times more massive than the Sun. It has probably lost about 10 Suns' worth of material so far over its lifetime of about 8 million years.

OUTPUT

VY Canis Majoris produces about 270,000 times more energy than the Sun.

TEMPERATURE

At about 3,200°C (5,800°F), the star's surface is much cooler than the Sun's.

INFRARED STAR

VY Canis Majoris gives off far more invisible infrared rays than visible light.

VARIABLE STAR

The star's brightness varies by 3 magnitudes over 956 days.

GLOWING SPHERE

VY Canis Majoris is depicted
as a seething red ball of glowing
gas in this artist's impression.
A strong wind of gas blows
constantly from its surface,
and active regions blast clouds
of material into space.

HEAVYWEIGHT STAR

R136a1

It would take about 9 million Suns, all shining together, to match the brilliance of R136a1 – the most massive, luminous star ever found. It is part of a magnificent star cluster within the Tarantula Nebula in the Large Magellanic Cloud, a galaxy near the Milky Way. Like R136a1, many of the stars in this cluster are Wolf-Rayet stars – rare, extremely hot, and unbelievably massive stars that have short lives and eject their gases at a phenomenal rate. R136a1 is less than two million years old but it is already halfway through its life.

AT A GLANCE

- **STAR TYPE** Wolf-Rayet star

- **DISTANCE FROM EARTH**
  About 163,000 light years

- **SURFACE TEMPERATURE**
  More than 49,000°C (88,232°F)

- **DIAMETER** About 45 million km
  (28 million miles)

R136a1 illuminates the central region of the Tarantula Nebula.

STATS AND FACTS

The rare group of Wolf-Rayet stars were first discovered in 1867 by French astronomers Charles Wolf and Georges Rayet. R136a1 itself was discovered in 2010 by astronomers using the Very Large Telescope in Chile and data gathered by the Hubble Space Telescope.

LIFESPAN

R136a1 is expected to last for only about 5 million years.

years (in million)   5,000   10,000

The Sun will last for 10,000 million years.

LARGE STARS

R136a1 is one of about 12 massive, luminous stars in the cluster R136.

SIZE

The star is 30 times wider than the Sun.

THE MOST MASSIVE, **LUMINOUS** STAR IN THE UNIVERSE

BRILLIANT LIGHT

This is an artist's impression of R136a1. A huge star like this emits a powerful bluish light. The discovery of R136a1 has made astronomers rethink how massive stars can get.

NEBULOUS HOME

Large billowing clouds of gas and dust in the Carina Nebula envelop the stars in the Eta Carinae system (the brightest spot in this image), hiding them from direct view.

UNSTABLE STARS

ETA CARINAE

Time is running out for the two stars that make up one of the most luminous, unstable star systems in the Milky Way – Eta Carinae. Combined, the stars are more than 5 million times brighter than the Sun, but when the most massive star in the system runs out of fuel, it will explode. Over the years, it has erupted a number of times and in the process, hurled a colossal cloud of dust and gas into space that is bigger than the Solar System. Astronomers expect the star system to explode in about a million years or even sooner. Meanwhile, the stars continue to release material.

AT A GLANCE

- **STAR TYPE** Variable star system

- **DISTANCE FROM EARTH**
  About 7,500 light years

- **DISCOVERED** In 1677 by English astronomer Edmond Halley

- **LOCATION** Within the Carina Nebula

  Eta Carinae is surrounded by the Homunculus Nebula – a dumbbell-shaped cloud of gas and dust – within the Carina Nebula.

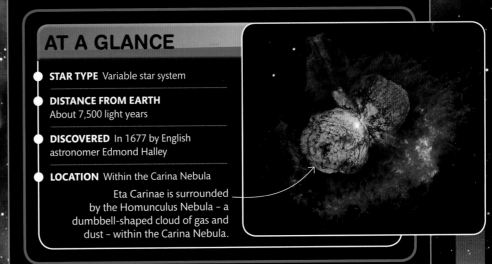

STATS AND FACTS

Between 1838 and 1843, Eta Carinae's brightness increased by more than 100 times. During that period, the star system released enough gas and dust to make ten Suns. The ejected material formed the Homunculus Nebula.

STAR MASS

The secondary star has a mass of about 50–80 times the mass of the Sun.

| 0 | 50 | 100 | 150 | 200 | 250 |

The primary star has a mass of about 150–200 times the mass of the Sun.

ORBIT

The two stars in the system orbit each other once every 5.5 years.

LASER LIGHT

Eta Carinae is the only known star to produce ultraviolet laser emissions.

MEGA STAR CLUSTER
OMEGA CENTAURI

Some of the oldest surviving stars in the Universe reside in dense clusters consisting of thousands or millions of stars. By far the brightest and largest of these globular clusters in the Milky Way galaxy is Omega Centauri. The stars are packed so tightly together that people once thought it was a single star, but the cluster contains up to 10 million stars held together by gravity – many nearly as old as the Universe. Omega Centauri may be the star-rich core of a dwarf galaxy that was stripped of its outer stars when it collided with the Milky Way.

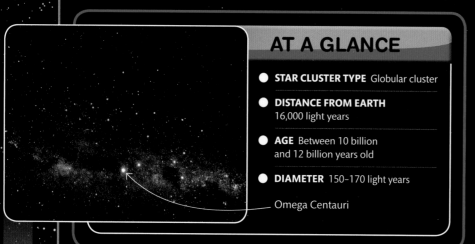

Omega Centauri

AT A GLANCE

- **STAR CLUSTER TYPE** Globular cluster

- **DISTANCE FROM EARTH**
  16,000 light years

- **AGE** Between 10 billion
  and 12 billion years old

- **DIAMETER** 150–170 light years

STATS AND FACTS

Ancient Greek scientist, Ptolemy described Omega Centauri as a single star in 150 CE. It was later labelled a nebula by the astronomer Edmond Halley in 1677, before finally being identified as a star cluster by the astronomer John Herschel in the 1830s.

STAR SEPARATION

light years

The average distance between stars in the core of Omega Centauri is only about 0.1 light years.

0   2.5   5

Proxima Centauri is about 4.2 light years away from its nearest star, Earth's Sun.

GLOBULAR CLUSTER

Omega Centauri is one of the 150 known globular clusters orbiting the Milky Way.

BLACK HOLE

A black hole more than 10,000 times as massive as the Sun might exist at the centre of the cluster.

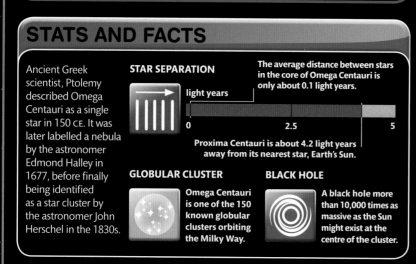

GLITTERING SIGHT

Omega Centauri is a spectacular sight in the skies of the southern hemisphere. Looking like a smudged star to the unaided eye, in a very dark sky it appears as large as the full Moon.

BLUE STARS
THE PLEIADES

Known since ancient times, the Pleiades is one of the closest, brightest star clusters to our planet, so it's fairly easy to view. The brilliant blue stars at its heart are just part of more than one thousand stars in the cluster. Their blue colour indicates a high temperature and young age – they are much hotter and more luminous than the Sun. The Pleiades is also called the Seven Sisters, after the daughters of Atlas and Pleione in Greek Mythology, and is known in Japan as Subaru. Like all open star clusters, the Pleiades will not last forever – they are loosely held together by gravity and are expected to drift apart over the next 250 million years.

AT A GLANCE

- **STAR CLUSTER TYPE** Open cluster

- **DISTANCE FROM EARTH**
  444 light years

- **DIAMETER** 86 light years

- **VISIBILITY** Best observed between November and January

The Pleiades star cluster is located near the constellation Orion.

STATS AND FACTS

A group of seven dots on the Nebra Sky Disc, an ancient bronze and gold artefact from 1,600 BCE, may represent the Pleiades. This disc is thought to be the world's oldest known depiction of the Universe.

AGE

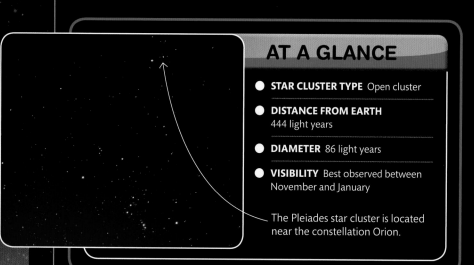

About 0.1 billion years old, the Pleiades is one of the youngest known star clusters.

years (in billion)   2.5   5

Our Sun is 4.5 billion years old.

BRIGHTEST STAR

The brightest star in the Pleiades, Alcyone, is 2,030 times brighter than the Sun.

OPEN CLUSTERS

The Pleiades is one of over 1,100 open star clusters within the Milky Way.

NIGHT-SKY JEWEL

The Pleiades star cluster is easily visible from Earth without using a telescope. It is embedded in a dust cloud that reflects the light of the brightest stars and makes it even more blue.

SPINNING STAR
PSR J1748-2446ad

Pulsars are small and incredibly dense stars that spin round at a furious rate while emitting beams of radiation from their magnetic poles. If these beams happen to sweep across Earth, we see rapid pulses of energy, just as a lighthouse appears to flash on and off as its beam sweeps past. Typical pulsars spin about once a second, but PSR J1748-244ad – the fastest pulsar known – rotates 716 times every second.

AT A GLANCE

PSR J1748-2446ad is in the star cluster Terzan 5.

- **STAR TYPE** Pulsar
- **DISTANCE FROM EARTH** 18,000 light years
- **DIAMETER** Less than 32 km (20 miles)
- **LOCATION** Constellation Sagittarius

Sweeping beams

A pulsar fires out narrow beams of electromagnetic radiation from its two opposite magnetic poles. These two beams travel through space, but astronomers can only detect the presence of a pulsar if one of the beams sweeps directly across Earth as the star spins.

The magnetic field is billions of times more powerful than Earth's magnetic field.

THE FASTEST SPINNING
PULSAR

STATS AND FACTS

42,960 RPM
REVOLUTIONS PER MINUTE

Pulsar PSR J1748-2446ad is located near the centre of the Milky Way.

NOT ALONE

PSR J1748-2446ad is one of 33 fast pulsars in Terzan 5.

SPIN

Pulsar PSR J1748-2446ad spins at a speed of more than 70,000 km/sec (43,000 miles/sec).

TEMPERATURE
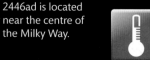
The surface temperature of this pulsar is about 600,000°C (1 million°F).

BINARY STAR

A companion star orbits PSR J1748-2446ad every 26 hours.

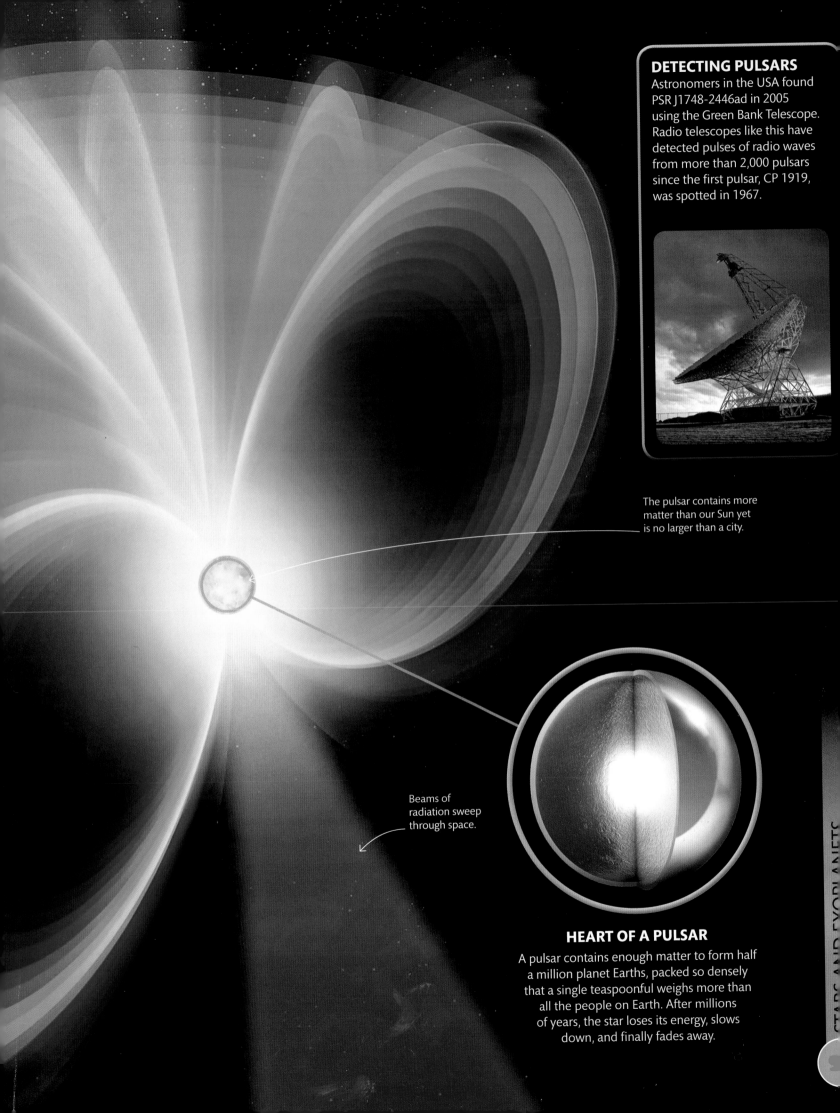

The pulsar contains more
matter than our Sun yet
is no larger than a city.

Beams of
radiation sweep
through space.

HEART OF A PULSAR

A pulsar contains enough matter to form half
a million planet Earths, packed so densely
that a single teaspoonful weighs more than
all the people on Earth. After millions
of years, the star loses its energy, slows
down, and finally fades away.

STARS AND EXOPLANETS

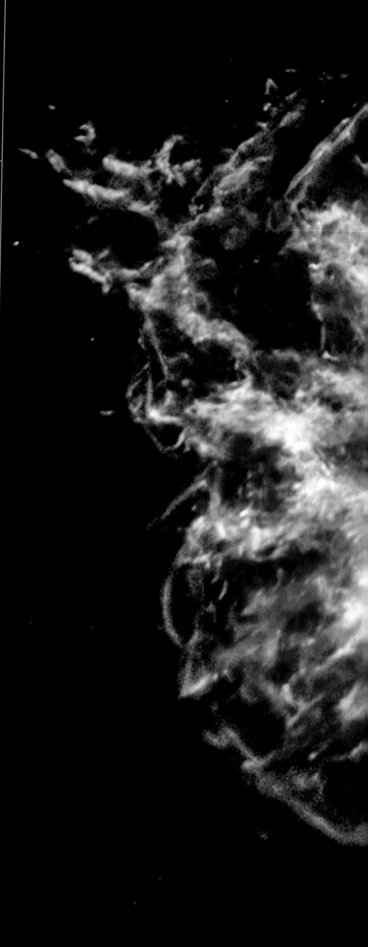

EXPLOSIVE END

CASSIOPEIA A

Stars end their lives in a number of ways but one of the most spectacular is known as a supernova – a colossal explosion. Cassiopeia A is all that remains of a massive star that was destroyed in such an explosion about 300 years ago. It has a rapidly growing shell of incredibly hot gas and dust. As it expands outwards, it disperses a huge variety of different chemical elements – most of the elements in your body come from supernovas like this. Cassiopeia A will continue to expand for thousands of years.

AT A GLANCE

- **STAR TYPE** Supernova remnant

- **DISTANCE FROM EARTH**
  11,000 light years

- **DIAMETER** 10 light years

- **DISCOVERED** In 1947 by British astronomers Martin Ryle and Francis Graham-Smith

Cassiopeia A is located in the W-shaped constellation Cassiopeia.

STATS AND FACTS

Cassiopeia A was discovered by its radio emission – it is one of the most intense radio sources in the sky. Thanks to spectacular images taken by Chandra, Spitzer, and Hubble, it provides astronomers with important clues as to how the star lived and died.

SPEED

	km/h (in million)	25		50		75
	mph (in million)		24			48

Jets of gas within the supernova remnant reach speeds of up to 50 million km/h (32 million mph).

STAR DEATH

When a massive star explodes, it briefly outshines an entire galaxy before fading away.

OXYGEN

The oxygen in the Solar System came from exploding massive stars like Cassiopeia A.

THE YOUNGEST SUPERNOVA REMNANT IN THE MILKY WAY.

COSMIC ELEMENTS

This colour-coded image of Cassiopeia A shows the presence of important elements, including calcium (green), iron (purple), silicon (red), and sulfur (yellow). The tiny white dot at the centre is a neutron star created by the explosion.

POWERFUL X-RAY SOURCE
SCORPIUS X-1

About 9,000 light years away from Earth is Scorpius X-1, a double star system that emits 100 million times more X-rays than the Sun. The interaction between a neutron star and a nearby donor star makes Scorpius X-1 the brightest source of X-rays outside the Solar System. Neutron stars form after massive stars die and their high density produces extreme gravity. This powerful force drags gases away from the donor star to make a rotating structure called an accretion disc around the neutron star. Friction between these gases in the disc generates temperatures of many millions of degrees, causing star systems such as Scorpius X-1 to release torrents of X-rays into space.

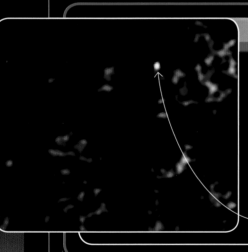

AT A GLANCE

- **STAR TYPE** Neutron star

- **DISCOVERED** In 1962 by American astronomer Riccardo Giacconi and colleagues at American Science and Engineering Inc

- **VISUAL MAGNITUDE** 11.8–12.7 (variable)

- **LOCATION** Constellation Scorpius

Scorpius X-1 is visible in this X-ray image taken by the Swift Telescope.

STATS AND FACTS

The first instruments used to detect X-rays from space were launched on small rockets that travelled to just outside Earth's atmosphere. Scorpius X-1 was detected in an attempt to track X-rays from the Moon.

X-RAY SOURCE

Scorpius X-1 was the first X-ray source found outside the Solar System.

X-RAY EMISSION

Scorpius X-1 emits over three times more X-rays than the next brightest X-ray source, the Crab Pulsar.

FREQUENCY

Some sources emit X-rays in bursts, but Scorpius X-1 emits them continuously.

OBSERVATION

NASA's Chandra X-ray Observatory has extensively studied this neutron star.

GAS RING

In this artist's impression of Scorpius X-1, gas from the donor star (shown in blue) is pulled by gravity into the large accretion disc (shown in red). The neutron star in the centre is encircled by these swirling gases and emits violent jets of heat and radiation above and below the disc.

SWIRLING CLOUDS

The central part of the Orion Nebula is a swirling mass of gas. This image from the Hubble Space Telescope has been made using artificial colours, which highlight the amazing patterns.

STAR-FORMING CLOUD

ORION NEBULA

Large numbers of stars are continually being created in the Orion Nebula, the nearest massive star-forming region to Earth. The huge clouds of dust and gas span an area over 1.5 million times wider than the distance between Earth and the Sun. Visible to the naked eye, the clouds are lit up by ultraviolet radiation from four young, hot stars. The stars are named the Trapezium because of the shape they form. The nebula is also home to many protostars (balls of hot gas that haven't become stars yet) and brown dwarfs (faintly glowing balls of gas that will never become proper stars).

AT A GLANCE

DISTANCE FROM EARTH
1,344 light years

DIAMETER 24 light years

BEST OBSERVED Between January and March

LOCATION Constellation Orion

The four stars that make up the Trapezium are among the largest in the Orion Nebula.

STATS AND FACTS

French astronomer Nicolas-Claude Fabri de Peiresc discovered the Orion Nebula in 1610. It was photographed for the first time in 1880 by American astronomer Henry Draper.

GAS EJECTION

Some infant stars in the Orion Nebula eject material travelling at speeds of up to 720,000 km/h (450,000 mph).

km/h (in 1,000)	500	1,000
mph (in 1,000)	310	620

NEW PLANETS

At least 150 discs, formed around stars in the nebula, may develop into future planets.

AGE

Scientists estimate that the nebula will last for another 100,000 years.

STAR NURSERY
CARINA NEBULA

Vast amounts of energy given off by super-hot newborn stars light up the billowing clouds of gas and dust that make up the Carina Nebula – one of the brightest, largest star-forming regions in our galaxy. Thousands of new stars form within this magnificent nebula, blowing off excess gas, dust, and charged particles as stellar winds. Rushing gases trigger turbulence in the clouds, which collapse and cause yet more stars to form. The nebula also contains spectacular star clusters like Trumpler 14, one of the youngest clusters in the Milky Way with at least 2,000 stars.

AT A GLANCE

- **AVERAGE DISTANCE FROM THE SUN**
  6,500–10,000 light years

- **DIAMETER** About 460 light years

- **DISCOVERED** In 1752 by French astronomer Nicolas-Louis de Lacaille

- **LOCATION** Constellation Carina

Carina Nebula

STATS AND FACTS

The vast Carina Nebula is home to both the smaller Keyhole and Homunculus nebulas. Some of the most massive stars in our galaxy, including the double star Eta Carinae, also reside within the Carina Nebula.

NUMBER OF STARS

The Chandra X-ray Observatory has detected about 14,000 stars within the Carina Nebula.

LUMINOUS STAR

Eta Carinae's total luminosity is over 5 million times greater than the Sun's.

SPEED

The Homunculus Nebula, the dusty cloud ejected by the star Eta Carinae, is expanding at more than 2.1 million km/h (1¼ million mph).

km/h (in million)	1.5	3
mph (in million)	1	2

SPACE WONDER

Taken by the Hubble Space Telescope, this image shows part of the Carina Nebula known as Mystic Mountain. Sculpted by stellar winds, the gas-dust tower is home to new stars. These fire off jets of gas, which will eventually erode the tower to reveal the once hidden stars.

RED EYE

This infrared image from the Spitzer Space Telescope shows the Helix Nebula. It looks like a giant red eye in space. The tiny dot just visible in the centre is the white dwarf star in the final stage of its life. Its disc of cold dust is nearly as wide as the Solar System and glows in infrared light.

NEIGHBOURING NEBULA

HELIX NEBULA

The Helix Nebula is one of the largest known planetary nebulas. It was formed when a dying star ran out of hydrogen fuel and cast away its layers of gas into space, creating this spectacular nebula. The star was originally similar to our Sun but, more than 10,000 years ago, as it drew towards the end of its life, it swelled up into a red giant. As the outer layers were thrown off, the hot core of the dying star was exposed and it illuminated the ejected debris. When gas from the star collided with denser gas nearby, clumps with streaming comet-like tails began to form. They were the first cometary knots ever recorded. There are thousands of them inside the Helix Nebula.

AT A GLANCE

- **DISTANCE FROM EARTH**
  Nearly 700 light years

- **DIAMETER** 2.5 light years

- **DISCOVERED** In 1823 by German astronomer Karl Ludwig Harding

- **LOCATION** Constellation Aquarius

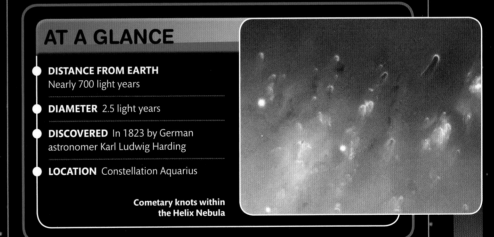

Cometary knots within
the Helix Nebula

STATS AND FACTS

The term planetary nebula originally came from early astronomers. When looking through their telescopes, they thought that these nebulas looked like the discs of planets.

TEMPERATURE

The temperature of the central star is about 100,000°C (180,000°F).

COMETARY KNOTS

The cometary knots in the Helix Nebula are as big as our Solar System.

SPEED

The outer layers of gas are moving at 144,000 km/h (90,000 mph).

km/h (in 1,000)	80	160
mph (in 1,000)	50	100

DISTANT WORLD
KEPLER-186f

Until relatively recently, the only known planets were those within our Solar System. But, with the help of more advanced telescopes, scientists have discovered thousands of planets orbiting stars far beyond our Solar System. These worlds, called exoplanets, range in size from giants twice as big as Jupiter to planets as small as our Moon. A few are part of a system of planets that resembles our Solar System, like Kepler-186f. It is one of five planets orbiting the same star, Kepler-186. Scientists think that these small worlds may have a rocky surface like Earth's.

AT A GLANCE

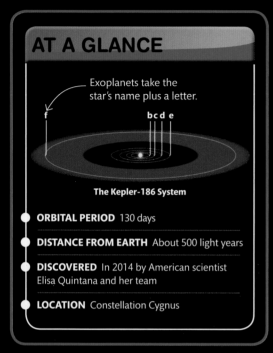

Exoplanets take the star's name plus a letter.

f   b c d e

The Kepler-186 System

ORBITAL PERIOD 130 days

DISTANCE FROM EARTH About 500 light years

DISCOVERED In 2014 by American scientist Elisa Quintana and her team

LOCATION Constellation Cygnus

Like Earth, Kepler-186f's rocky surface might have liquid water.

Earth-like planet

This artist's impression shows Kepler-186f with possible Earth-like features. It was detected in 2014 by NASA's Kepler Space Telescope and lies near the outer edge of a belt around its star, known as the habitable zone, in which the conditions could be suitable for liquid water to exist and for life to flourish. The other new-found worlds in this system, Kepler-186 b, c, and d, are very close to their star, making it too hot for life to exist there.

Kepler-186 is a red dwarf star about half the size of the Sun.

KEPLER-62 SYSTEM

These artist's impressions show a five-planet system, which is orbiting a star 990 light years from Earth and was discovered in 2013 by the Kepler Space Telescope. It has a Mars-sized planet (Kepler-62c) with a surface temperature of 300°C (572°F) and two massive rocky planets (62e and 62f) up to ten times bigger than Earth. Both these exoplanets have surface temperatures potentially suitable for life to exist.

KEPLER-62b

KEPLER-62c

KEPLER-62d

KEPLER-62e

KEPLER-62f

"Kepler-186f was the first **Earth-sized** exoplanet discovered in a star's habitable zone."

STATS AND FACTS

Exoplanets are hard to detect because they are small and hidden by the glare of the star they orbit. The first ones were detected in the 1990s, and the Kepler Space Telescope has since found thousands. Scientists think there may still be more undiscovered planets in the Kepler-186 System.

STAR DISTANCE

Kepler-186f orbits 65 million km (40 million miles) from its sun.

km (in million)	80	160
miles (in million)	50	100

Earth orbits 150 million km (93 million miles) from its Sun.

DIAMETER

The diameter of Kepler-186f is about 15,000 km (9,000 miles), which is slightly bigger than Earth.

AGE

Scientists think that the Kepler-186 System is about 4 billion years old.

ALIEN WORLD

This artist's impression shows the star Proxima
Centauri looming over the imagined rocky
landscape of the exoplanet Proxima b. While
Earth is the only place where life is known
to exist, astronomers scour the cosmos in
their hunt for an exoplanet lying within a star's
habitable zone, where the temperature is
right for liquid water and life could flourish.

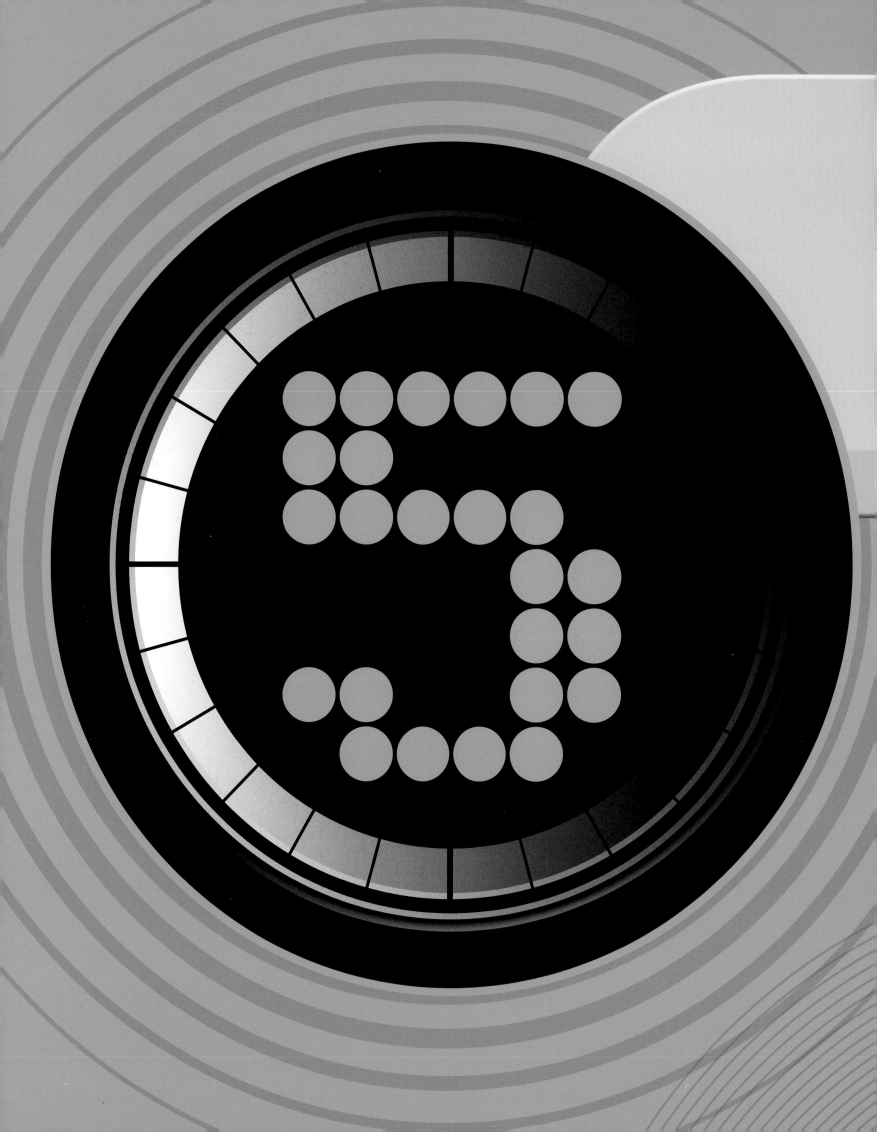

GALAXIES

The Universe contains more than 125 billion galaxies. Each one is a collection of countless stars, gas, and dust clouds, all bound together by gravity. Galaxies first formed billions of years ago when groups of stars merged together, creating the spectacular variety of shapes and sizes seen today.

OUR GALACTIC HOME
THE MILKY WAY

Look up on a dark, clear night and you might spot a band of misty light running across the sky, spangled with stars and nebulas. This is the Milky Way – the galaxy our Solar System belongs to. We see it as a band because we're inside it, but it is a gigantic whirlpool of more than 200 billion stars. Vast arms of dust, gas, and stars spiral out from a bar-shaped bulge in the centre. In the very heart of the galaxy lies a supermassive black hole.

AT A GLANCE

Hubble image of the centre of the Milky Way

- **GALAXY TYPE** Barred spiral

- **LOCATION** The centre of the Milky Way is about 25,500 light years from the Solar System

- **AGE** More than 13 billion years old

- **DISCOVERED** Ancient times; both the Greeks and Romans compared the Milky Way to spilt milk

Sagittarius Arm

Perseus Arm

Outer Arm

Spiral arms

Viewed face-on, as in this artist's impression, the disc of our home galaxy would look similar to other barred spiral galaxies. Several spiral arms wind out from the central bar. Stars in the arms do not stay there permanently but, moving in orbit around the centre, they pass through the arms like vehicles through a traffic jam.

STATS AND FACTS

People once thought that the Milky Way galaxy was the whole of the Universe. This changed in 1923 when US astronomer Edwin Hubble proved the existence of other galaxies by showing that some spiral-shaped structures were beyond the Milky Way.

ORBIT

The Solar System orbits the centre of the Milky Way at 864,000 km/h (537,000 mph).

km/h (in 1,000)	500	1,000
mph (in 1,000)	310	620

HUGE HALO

A giant halo of gas surrounds the Milky Way and extends thousands of light years into space.

STAR MAP

The Gaia spacecraft took five years to map one billion stars in the Milky Way.

SIDE VIEW

Viewed from the side, the Milky Way resembles two fried eggs placed back-to-back. The galaxy's main disc is about 1,000 light years thick but the large bulge in its centre is six times thicker. Above and below the bulge and disc are globular clusters – balls of very ancient stars.

Main disc containing spiral arms

Barred central bulge packed with millions of stars

Norma Arm

Scutum-Centaurus Arm

Glowing clouds of hydrogen gas where new stars are forming

Orion Spur

OUR PLACE IN SPACE

Our Solar System and Earth are located on the inner edge of the Orion Spur, a minor arm about 25,000 light years long lying between the Perseus and Sagittarius arms. All of the brightest stars in our night sky lie in this local part of the galaxy.

The central bulge is about 27,000 light years long.

Globular cluster of stars above central bulge

GALAXIES

HOME GALAXY

On a clear night, if the sky is very dark, you have a good chance of seeing the Milky Way. This glowing, hazy band across the sky is the light of billions of distant stars belonging to our galaxy. Some of the light is blocked out by the dark, dusty clouds within the galaxy.

NO ESCAPE
SAGITTARIUS A*

Probably the most mysterious and mind-boggling object in the Universe is a black hole. This is a place where a vast amount of matter is being crushed into a point that is even tinier than an atom. The pulling force of a black hole's gravity is so strong that nothing can escape it – anything that gets too close is dragged in, from planets and stars to vast clouds of interstellar gas and dust. Not even light can escape its grasp, making the black hole invisible. Space telescopes have peered through the dust- and gas-filled clouds of our own galactic home, the Milky Way, and found evidence of a supermassive black hole known as Sagittarius A* in the very centre of the galaxy.

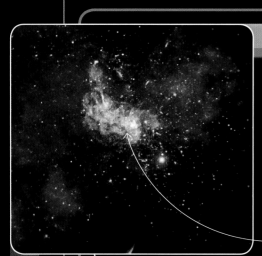

AT A GLANCE

● **DISTANCE FROM EARTH** About 26,000 light years

● **MASS** About 4 million Suns

● **DISCOVERED** In 1974 by American astronomers Bruce Balick and Robert Brown

● **LOCATION** Constellation Sagittarius

Sagittarius A* lies within this bright region of the Milky Way.

STATS AND FACTS

Most galaxies have supermassive black holes at their centres, but they are too far away to study. Sagittarius A* is one of the very few in the Universe where scientists can observe the flow of matter.

SUPERFAST STARS

Stars near Sagittarius A* orbit the black hole at speeds up to 5 million km/h (3 million mph).

MORE BLACK HOLES

Data from Chandra has revealed thousands of smaller black holes around the supermassive one.

TIME

The black hole's gravity distorts time – it runs slower near the black hole.

GRAVITATIONAL PULL

Matter falling into the black hole gets stretched into a spaghetti-like shape.

WHIRLPOOL

This artist's impression shows a supermassive black hole at the centre of a galaxy. It is surrounded by a disc of hot, glowing material spiralling into the black hole. Anything that passes too close to the black hole is pulled in and trapped forever.

GALACTIC NEIGHBOUR
ANDROMEDA GALAXY

THE CLOSEST LARGE GALAXY TO EARTH

The magnificent Andromeda Galaxy is the most distant object visible with the naked eye from Earth. It was formed about 10 billion years ago from several smaller galaxies, and contains a trillion stars – more than twice the number of our own galaxy the Milky Way. Andromeda also has about 460 globular clusters, each containing hundreds of thousands of stars, and is orbited by a number of much smaller galaxies, held by its powerful gravity. In about 4 billion years, Andromeda is expected to collide with the Milky Way, forming a giant galaxy.

AT A GLANCE

- **GALAXY TYPE** Spiral

- **DISTANCE FROM EARTH**
  2.54 million light years

- **MAGNITUDE** 3.4

- **DISCOVERED** First recorded mention of Andromeda was by Arabic astronomer Abd al-Rahman al-Sufi in 964 CE in his *Book of Fixed Stars*

Andromeda appears as an oval smudge next to the Milky Way.

STATS AND FACTS

In 1925, American astronomer Edwin Hubble proved that Andromeda was a separate galaxy and not just a gas cloud within the Milky Way as was originally thought. Hubble went on to devise a system of classifying galaxies.

DIAMETER

light years (in 1,000)

The Andromeda Galaxy is 220,000 light years in diameter.

150   300

The Milky Way is just 100,000 light years across.

COLLISION COURSE

Andromeda is hurtling towards the Milky Way at 400,000 km/h (248,550 mph).

BLACK HOLE

The galaxy's black hole is several hundred million times more massive than the Sun.

SPIRAL GIANT

This image reveals the beautiful natural colour of Andromeda. The dense bright core consists of old stars while the spiral arms are full of luminous young stars. A number of smaller galaxies circle Andromeda.

SPECTACULAR STARBURST

M82

The dazzling M82 galaxy shines five times brighter than our Milky Way. It is seen in the constellation Ursa Major, and is a starburst galaxy. It looks so spectacular because stars are forming at a rapid rate in its core. M82 is only a third of the size of the Milky Way, but the number of new stars being born is ten times higher. The first burst of star formation occurred about 600 million years ago when M82 almost collided with its galactic neighbour, M81. The full force of M81's gravitational pull disturbed M82's galactic core and triggered the formation of numerous new star clusters.

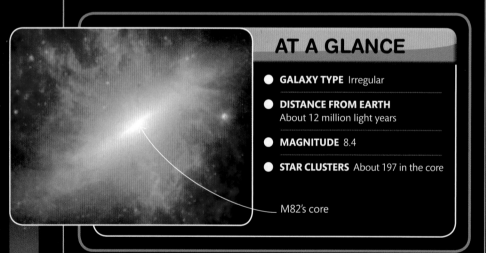

M82's core

AT A GLANCE

- **GALAXY TYPE** Irregular
- **DISTANCE FROM EARTH**
  About 12 million light years
- **MAGNITUDE** 8.4
- **STAR CLUSTERS** About 197 in the core

STATS AND FACTS

In 1774, German astronomer Johann Elert Bode discovered M82. Unaware of his work, French astronomer Pierre Méchain discovered this galaxy again in 1779.

DIAMETER

M82 is 37,000 light years across.

| light years (in 1,000) | 40 | 80 | 120 |

Our galaxy, the Milky Way, is about 100,000 light years across.

BRIGHT CORE

M82's core shines almost 100 times more brightly than the Milky Way's core.

BRIGHTEST PULSAR

The galaxy contains M82 X-2, one of the brightest pulsars in the Milky Way.

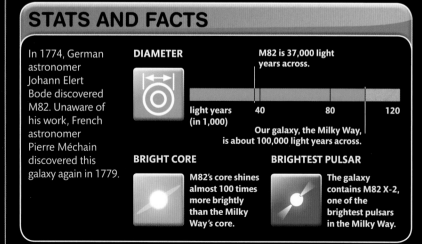

The nearest starburst galaxy to the Milky Way, M82 is relatively small and surrounded by swirling gas and dust. This stunning image, taken by the Hubble Space Telescope, shows visible and infrared wavelengths together with the red glow of hydrogen released from intense starbursts.

THE FIRST
GALAXY FOUND
TO BE SPIRAL

SPIRAL
WHIRLPOOL GALAXY

The spiral arms of the Whirlpool Galaxy are home to billions of stars. The long, dark lanes of gas and dust create the effect of a cosmic whirlpool. As in most galaxies, a supermassive black hole resides at the centre. The galaxy, also known as M51a, is similar in size to our Milky Way, but it has expanded by interacting with a neighbouring irregular galaxy named M51b. Millions of years ago, gravity from this dwarf galaxy began pulling on the spiral arms of the Whirlpool. This action compressed hydrogen gas to create clusters of new stars.

AT A GLANCE

GALAXY TYPE Spiral

DISTANCE FROM EARTH
23 million light years

MAGNITUDE 8.4

DISCOVERED In 1773 by French astronomer Charles Messier

The Whirlpool Galaxy's bright core is visible in this photo taken by the Hubble Space Telescope.

STATS AND FACTS

The Whirlpool Galaxy's shape was identified in 1845 when Irish astronomer Lord Rosse observed its spiral arms using his reflecting telescope.

NUMBER OF STARS

100 BILLION

VISIBILITY

The Whirlpool Galaxy is visible with binoculars in the constellation Canes Venatici.

SUPERNOVA

Supernovas are rare but three were spotted in the galaxy between 1994 and 2011.

DIAMETER

The galaxy has a diameter of about 65,000–80,000 light years.

INTERACTING GALAXIES

A bridge-like connection of gas and dust links the Whirlpool with M51b.

LUMINOUS GALAXY

Light from newly formed stars illuminates the Whirlpool's spiral arms in this photograph taken by the Hubble Space Telescope. The youngest stars are found in the bright outlying regions, while the oldest stars are in the centre. Galaxy M51b is visible on the upper right, passing slowly behind one of M51a's arms where new stars are forming.

COSMIC COLLISION
ANTENNAE GALAXIES

Pulled towards each other by their own gravity, the galaxies NGC 4038 and NGC 4039 have been crashing into one another for several hundred million years. As the galaxies collide, gas and dust become compressed by the impact and create millions of new stars. Two long strands of gas, dust, and stars trail from the cores of the two galaxies. They resemble an insect's antennae, leading to the name by which the galaxies are commonly known. The trails, formed when the galaxies first collided, show what may happen when the Milky Way collides with the Andromeda Galaxy in about four billion years.

AT A GLANCE

- **GALAXY TYPE** Barred spiral (NGC 4038) and spiral (NGC 4039)

- **DISTANCE FROM EARTH** 60–65 million light years

- **FIRST ENCOUNTER OF THE GALAXIES** About 500 million years ago

- **DISCOVERED** In 1785 by German-British astronomer William Herschel

STATS AND FACTS

About a billion years ago, the Antennae Galaxies were separate galaxies. In another 300–400 million years, their cores will have merged to form one single galaxy.

NUMBER OF STARS

The colliding galaxies are thought to contain nearly 300 billion stars.

CORES

The distance between the cores of the two galaxies is about 30,000 light years.

STAR LIFE

Most of the new stars in the Antennae Galaxies will only last about 10 million years.

WIDTH

The Antennae Galaxies are five times wider than the Milky Way.

COLOURFUL CHAOS

In this image from the Hubble Space Telescope, vast clouds of gas appear pink and red, while star-forming regions glow blue. The dark lines are dust blocking some of the light, and the two bright orange blobs are the cores of the original galaxies.

COSMIC POWERHOUSE
ACTIVE GALAXY ULAS J1120+0641

Active galaxies in the distant Universe are among the brightest and most energetic objects. These galaxies produce huge amounts of radiation from their cores rather than their stars. The brilliant galaxy ULAS J1120+0641 is a quasar, the most violent type of active galaxy. Its dazzling brightness comes from the enormous energy released by the supermassive black hole in its centre. As a hot spinning disc of dust and gas is pulled constantly into the black hole, incredible amounts of high-energy radiation are emitted into space.

Brilliant quasar

An artist's impression of ULAS J1120+0641 shows how an active galaxy is similar to any other galaxy, but with an extremely bright centre and powerful particle jets shooting into space. This quasar formed about 13 billion years ago.

Rings of thick gas and dust build up around the centre of the galaxy and block views further inside.

A supermassive black hole lies at the centre of the quasar.

AT A GLANCE

ULAS J1120+0641 is the most distant known quasar and appears as a faint red dot in this image.

GALAXY TYPE Quasar

DISTANCE FROM EARTH 29 billion light years

DISCOVERED In 2011 by the UKIRT Infrared Deep Sky Survey

LOCATION Constellation Leo

"ULAS J1120+0641 shines as brightly as 63 trillion Suns."

Jets of electrically charged particles blast out of the galaxy's centre at right angles to the accretion disc.

Gas and objects torn apart by the black hole's gravity form an accretion disc that spirals around the edge of the black hole.

TYPES OF ACTIVE GALAXY

All active galaxies have energetic cores. Astronomers divide them up into four different groups depending on their activity levels and how they look when viewed from Earth.

Radio galaxy

This typical radio galaxy, Pictor A, emits radio waves from gigantic lobes of gas that are powered by particle jets more than 300,000 light years long, fired out of the active centre.

Seyfert galaxy

Seyfert galaxies are typically spirals, such as NGC 6814 pictured here, but their centres are exceptionally bright. NGC 6814 also emits X-rays.

Quasar

A quasar is the most distant type of active galaxy and has an extremely bright and active nucleus. This artist's impression shows light blazing from the central disc of SDSS J1106+1939.

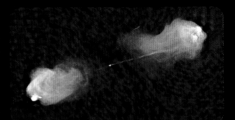

Blazar

Active galaxy Cygnus A shoots powerful jets across the sky but when such jets are directed straight at Earth, the galaxy is called a blazar.

STATS AND FACTS

Active galaxies produce different types of radiation from the central core, including radio waves, X-rays, and gamma rays. Two giant jets measuring many light years long also blast out from the axis of the black hole.

DISTANT QUASAR

ULAS J1120+0641 is the second most distant quasar discovered.

BLACK HOLE

The black hole at the centre of ULAS J1120+0641 is two billion times more massive than our Sun.

SPEEDY JETS

Jets emitted from quasars can travel at almost the speed of light.

BRIGHT QUASAR

Some quasars emit 10,000 times more energy than the entire Milky Way.

GALACTIC HAT
SOMBRERO GALAXY

The Sombrero Galaxy takes its name from the Mexican hat because of its wide brim and bulging centre. This galaxy is a spiral like our Milky Way, but the Sombrero's swirling arms are hard to distinguish against its blazing, bright core. That dazzling glow in the central bulge comes from about 2,000 clusters of ancient stars, ten times more than in the Milky Way. Altogether, there are about 1.3 trillion stars, including the ones in the outer disc. Observations suggest the Sombrero's supermassive black hole could be one of the most massive found in any nearby galaxy.

AT A GLANCE

- **GALAXY TYPE** Spiral
- **DISTANCE FROM EARTH** 29.3 million light years
- **DISCOVERED** In 1781 by French astronomer Pierre Méchain
- **LOCATION** Constellation Virgo

Sombrero Galaxy

STATS AND FACTS

Although this galaxy can be spotted with binoculars, a small amateur telescope is needed to see the disc clearly, and a more powerful one to detect the dust lane.

SPEED

The galaxy is moving away from us at a speed of 1,024 km/sec (636 miles/sec).

km/sec	500	1000	1500
miles/sec		465	930

DIAMETER

50,000 LIGHT YEARS

SIZE

The Sombrero Galaxy is about half as wide as the Milky Way Galaxy.

ANCIENT CLUSTERS

The clusters of ancient stars in this galaxy are 10-13 billion years old.

DUST RING

This image taken by the Hubble Space Telescope shows the Sombrero's glowing halo. A broad, dark ring of dust and hydrogen gas encircles the galaxy, blocking some of the starlight and forming the "brim" of the Sombrero.

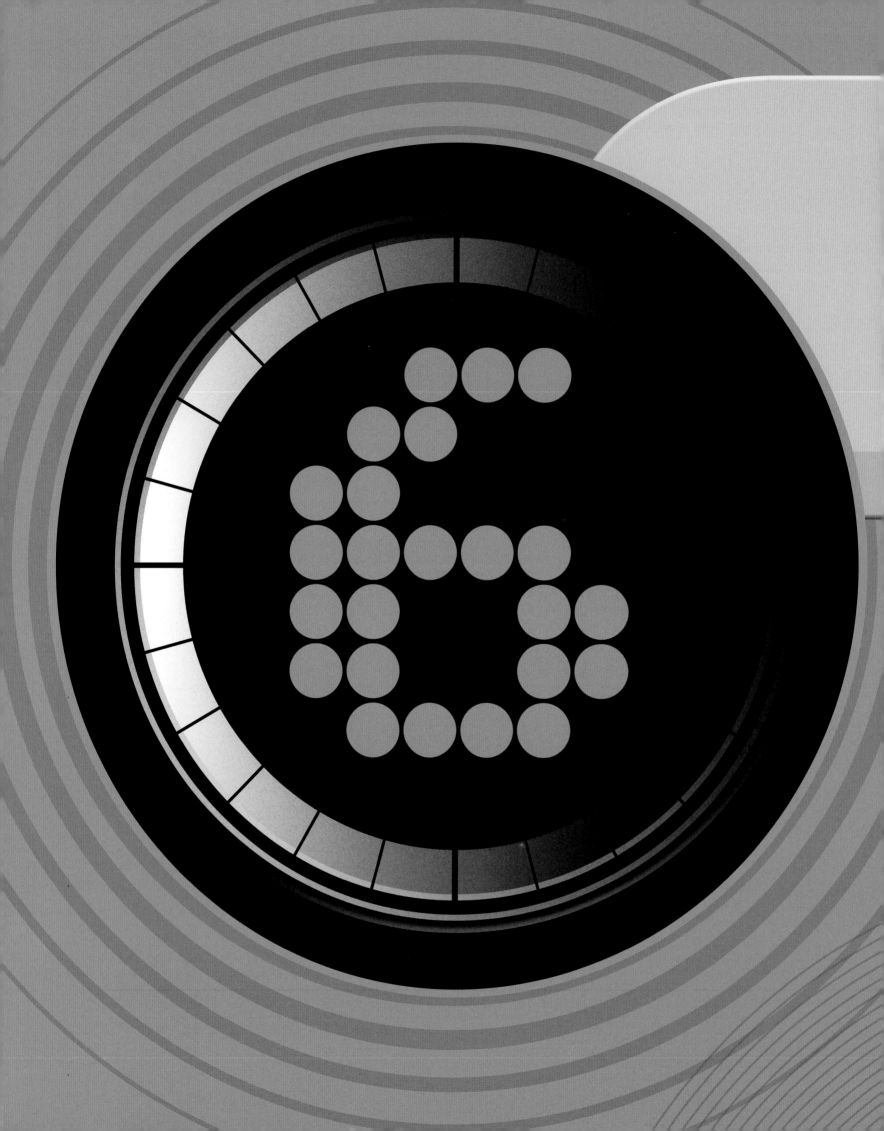

SPACE EXPLORATION

People have been fascinated by the stars and planets since the earliest times. After telescopes were invented in the 1600s, astronomers began to discover new planets, moons, and stars. Today, robotic spacecraft travel through the Solar System to fly by and even land on the planets.

CUTTING-EDGE DESIGN

With their huge segmented mirrors and a host of sensitive electronics, the Keck telescopes have astonishing observing powers. They bring into view space objects that are more than 10 billion light years away.

WINDOW TO SPACE
KECK TELESCOPES

The Keck Observatory in Hawaii houses two of the world's largest optical telescopes. Perched near the summit of a dormant volcano, the observatory is located far from city lights under a night sky with few clouds, which makes it ideal for observing space. The main mirrors in these telescopes are larger than most other light-collecting mirrors on Earth. In more than 20 years of operation, these telescopes have allowed scientists to see distant galaxies, exoplanets, and the region around the supermassive black hole at the centre of our Milky Way galaxy.

AT A GLANCE

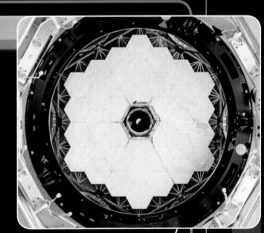

- **TELESCOPE TYPE** Optical/infrared telescopes

- **LOCATION** Mauna Kea, Hawaii

- **OPERATIONAL** 1993 (Keck I), 1996 (Keck II)

- **ORGANIZATION** California Association for Research in Astronomy

The main mirror of each Keck telescope is about as wide as a tennis court.

STATS AND FACTS

Earth's atmosphere blurs our view of space objects. The Keck telescopes solve this problem by using a system that corrects any distortion by moving special small mirrors as often as 2,000 times a second to obtain sharp images.

MIRROR SIZE

Each main mirror has a diameter of 10 m (32¾ ft).

SEGMENTED MIRROR

Thirty-six hexagonal mirrors controlled by computer act like a single mirror.

IMAGE QUALITY

Keck images can be as sharp as the Hubble Space Telescope's.

EXOPLANET HUNTER

In 2017 alone, the Keck telescopes discovered more than 100 possible exoplanets.

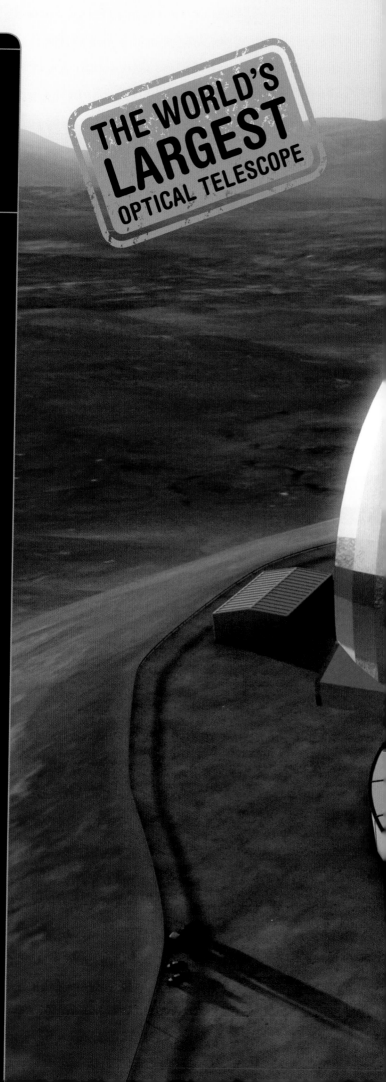

SPACE SCANNER
EXTREMELY LARGE TELESCOPE

On a mountain site in Chile, engineers are currently building the world's largest telescope. It will be 100 million times more sensitive than the human eye, allowing it to peer deeper into space than any optical telescope before it. Set in a dome as high as a 15-storey building, the telescope will scan the skies for supermassive black holes, observe stars being born, and even look for habitable planets and the first galaxies. Viewing the Universe in so much greater detail than ever before will enable scientists to make groundbreaking discoveries.

AT A GLANCE

- **TELESCOPE TYPE**
  Optical/Infrared telescope

- **LOCATION** Atacama Desert, northern Chile

- **OPERATIONAL** 2024

- **ORGANIZATION** European Southern Observatory (ESO)

This is the ELT's main mirror.

STATS AND FACTS

Once operational, the telescope will use a special moving mirror and laser beams fired into the sky to make its images very sharp. The building housing the telescope will also have spring-like devices to stop vibrations from earthquakes shaking the telescope.

WEIGHT

The ELT will weigh more than 240 times the weight of the Hubble Space Telescope.

ALTITUDE

The ELT is located at a height of more than 3 km (1¾ miles) above sea level.

PRIMARY MIRROR

The main mirror will have an area of 978 sq m (10,527 sq ft) – more than 3½ times the size of a tennis court.

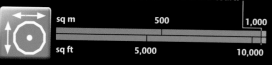

sq m		500		1,000
sq ft		5,000		10,000

"The telescope's main mirror is made up of 798 hexagonal segments."

EYE ON THE SKY

The telescope is being built in an area of northern Chile that has cloud-free skies and is away from light interference from towns. The telescope's dome is shown opening in this artist's impression.

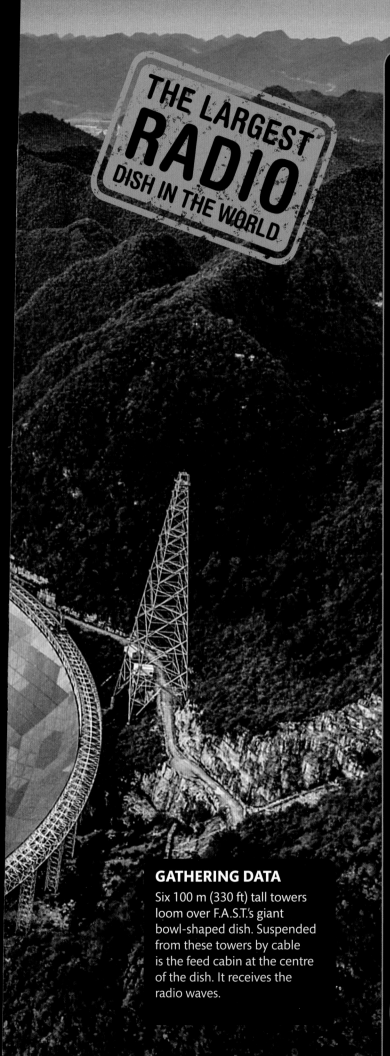

THE LARGEST RADIO DISH IN THE WORLD

SPACE RADIO
F.A.S.T.

The world's largest single-dish radio telescope sits in a natural basin in the remote mountains of China. The aptly named Five-hundred-meter Aperture Spherical Telescope (F.A.S.T.) collects faint radio waves emitted by objects such as galaxies and pulsars, two of which it discovered within weeks of starting observations in 2017. The surface of the dish is made up of 4,450 triangular aluminium panels, all fitting together like a giant jigsaw. More than 2,000 computer-controlled winches pull on these panels to enable parts of the dish to focus on a particular target in space.

AT A GLANCE

- **TELESCOPE TYPE** Single dish radio telescope
- **LOCATION** Pingtang County, southwest China
- **BUILT** 2011–2016
- **ORGANIZATION** National Astronomical Observatories of China

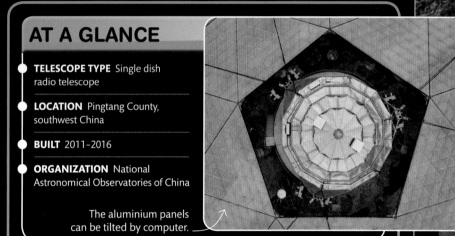

The aluminium panels can be tilted by computer.

GATHERING DATA

Six 100 m (330 ft) tall towers loom over F.A.S.T.'s giant bowl-shaped dish. Suspended from these towers by cable is the feed cabin at the centre of the dish. It receives the radio waves.

STATS AND FACTS

F.A.S.T. can detect very faint cosmic radio sources because its dish is so large. It is expected to make many new discoveries when fully working.

DIAMETER

F.A.S.T.'s dish is 500 m (1,640 ft) in diameter.

m	250	500	750
ft	1,230		2,460

The Arecibo Observatory in Puerto Rico has a 305 m (1,000 ft) wide dish.

DISH AREA

ABOUT **30** FOOTBALL PITCHES

DATA

F.A.S.T. can gather an amazing 3 million gigabytes of data every year.

SEARCH FOR LIFE

As well as looking for pulsars, F.A.S.T. is scouring space for possible alien radio signals.

Sitting on a pedestal structure, each dish can turn and tilt. All of them point at the object the telescope is observing. They can also be moved along tracks to alter the size of the array.

THE MOST ADVANCED **RADIO** TELESCOPE ARRAY

WORKING TOGETHER
VERY LARGE ARRAY

Radio astronomers have two choices when they want to pick up faint radio waves from space: they can either build a big dish antenna or construct lots of smaller antennae that work together as an array. One of the most powerful and versatile radio telescopes in the world is called the Very Large Array (VLA). Each of the 27 dishes is like a massive satellite dish and contains eight receivers that detect radio waves emitted by stars and other space objects. Quasars, black holes, distant galaxies, and even ice on Mercury have all been discovered by the VLA.

AT A GLANCE

- **TELESCOPE TYPE** Radio telescope
- **LOCATION** San Agustin plains, New Mexico, USA
- **BUILT** 1973–1980
- **ORGANIZATION** National Radio Astronomy Observatory (NRAO)

The antennae are arranged in a Y shape.

STATS AND FACTS

In 2012, the VLA was named after Karl G Jansky who built an early radio telescope in 1933. It could turn round, perched on tyres from a Model T Ford car.

MIRROR SIZE

Each dish in the VLA is 25 m (82 ft) in diameter.

m	15	30	45
ft	75		150

SUPERCOMPUTER

The computer that analyses data from the VLA can perform 10,000 trillion operations every second.

(in trillion)	6,000	12,000

WEIGHT OF DISH

209 TONNES

UNIVERSAL OBSERVER

HUBBLE SPACE TELESCOPE

No other telescope has contributed to as many different astronomical discoveries as the Hubble Space Telescope, from our Solar System to the most distant galaxies. It has made more than 1.3 million separate observations of objects in space. Orbiting above the atmosphere, this automated observatory can detect ultraviolet and infrared light that cannot get through to telescopes on the ground and its images are pin-sharp, untroubled by the blurring effect of Earth's atmosphere.

SECONDARY MIRROR

This smaller mirror reflects light from the primary mirror onto the telescope's cameras and instruments. It is made from glass coated in a thin layer of aluminium and kept at a constant temperature of 21°C (70°F) to stop it losing shape.

AT A GLANCE

TELESCOPE TYPE Optical/Ultraviolet/Infrared telescope

LAUNCH DATE 24 April 1990; the Hubble was flown into space by the Space Shuttle Discovery

MISSION DURATION Ongoing

ORGANIZATION NASA

Solar panels produce electricity, which is stored by six batteries inside the telescope.

The aperture door is closed to protect the telescope's optics during servicing missions by astronauts.

Light enters the telescope through this aperture and bounces off mirrors inside.

Grand design

The Hubble Space Telescope is about the same size as a bus. It orbits Earth at a speed of 27,400 km/h (17,000 mph), detecting optical, ultraviolet, and infrared wavelengths from objects in space. The scientific instruments and mirrors are held firmly in place by a lightweight graphite frame, and the telescope's outer shell of aluminium is covered in heat-reflecting foil. Astronomers from many different nations operate Hubble by remote control.

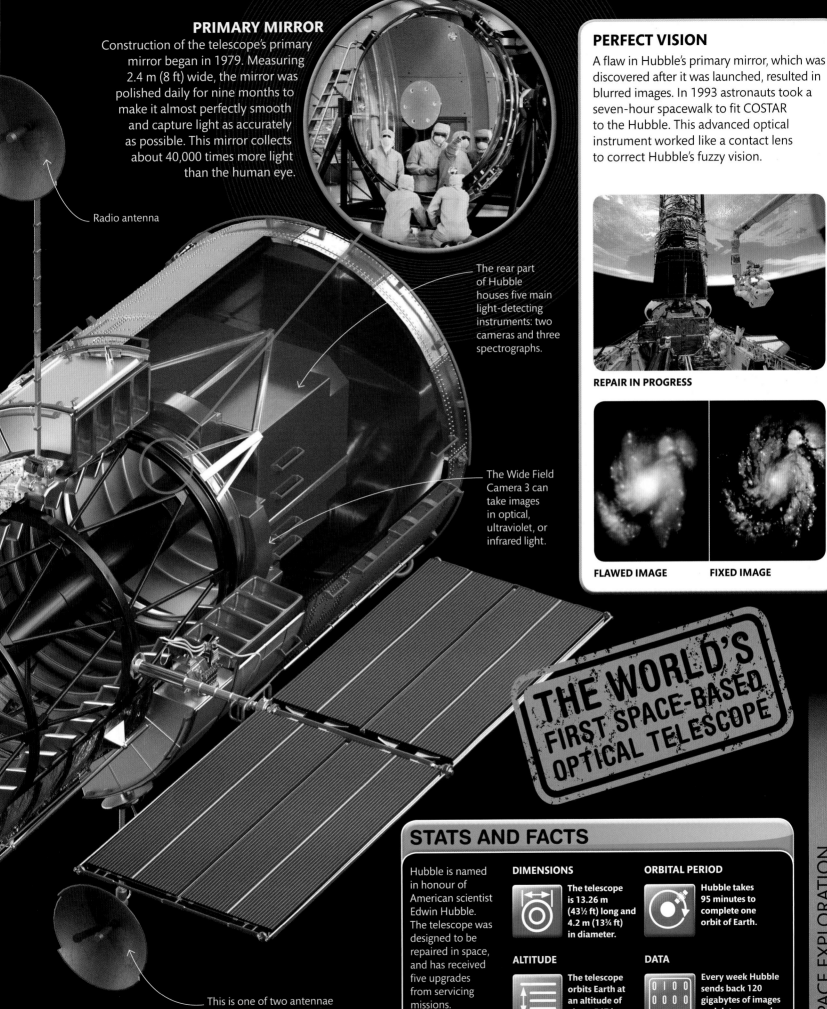

PRIMARY MIRROR

Construction of the telescope's primary mirror began in 1979. Measuring 2.4 m (8 ft) wide, the mirror was polished daily for nine months to make it almost perfectly smooth and capture light as accurately as possible. This mirror collects about 40,000 times more light than the human eye.

Radio antenna

PERFECT VISION

A flaw in Hubble's primary mirror, which was discovered after it was launched, resulted in blurred images. In 1993 astronauts took a seven-hour spacewalk to fit COSTAR to the Hubble. This advanced optical instrument worked like a contact lens to correct Hubble's fuzzy vision.

REPAIR IN PROGRESS

FLAWED IMAGE   **FIXED IMAGE**

The rear part of Hubble houses five main light-detecting instruments: two cameras and three spectrographs.

The Wide Field Camera 3 can take images in optical, ultraviolet, or infrared light.

This is one of two antennae that receive instructions from Earth and send data back for scientists to study.

THE WORLD'S FIRST SPACE-BASED OPTICAL TELESCOPE

STATS AND FACTS

Hubble is named in honour of American scientist Edwin Hubble. The telescope was designed to be repaired in space, and has received five upgrades from servicing missions.

DIMENSIONS

The telescope is 13.26 m (43½ ft) long and 4.2 m (13¾ ft) in diameter.

ALTITUDE

The telescope orbits Earth at an altitude of about 547 km (340 miles).

ORBITAL PERIOD

Hubble takes 95 minutes to complete one orbit of Earth.

DATA

Every week Hubble sends back 120 gigabytes of images and data – enough to fill 25 DVDs.

LOOKING AT THE UNIVERSE

In 2018, the Hubble Space Telescope captured this stunning view of the clouds of gas in the Lagoon Nebula, which lies more than 4,000 light years from Earth. The nebula is so large and luminous that it is even visible to the naked eye in the night sky.

X-RAY VISION

CHANDRA X-RAY OBSERVATORY

On an elongated orbit far above Earth's atmosphere, up to 240 times higher than the Hubble Space Telescope – Chandra detects X-rays emitted from the hottest objects in space. It has helped scientists peer into the Universe, shedding light on the behaviour of dying stars and colliding galaxies, and has even watched a super massive galaxy swallow a neighbouring one. Its mission was initially expected to last five years but because of the unprecedented details it sends back, the Chandra X-ray Observatory continues exploring the hidden Universe.

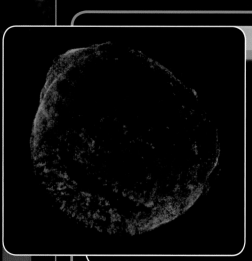

AT A GLANCE

- **TELESCOPE TYPE**
  X-ray space telescope
- **ORBITAL PERIOD** Takes about 64 hours to orbit Earth
- **LAUNCH DATE** 23 July 1999
- **ORGANIZATION** NASA and Smithsonian Astrophysical Observatory (SAO)

Chandra's X-ray image of the remains of a star that exploded more than 1,000 years ago.

STATS AND FACTS

The X-ray telescope was named after the Nobel Prize-winning astrophysicist Subrahmanyan Chandrasekhar. It is one of NASA's four powerful space-based telescopes designed to observe the Universe at different wavelengths.

LENGTH

At 13.8 m (45⅓ ft), Chandra is the longest space telescope launched so far.

m		5		10		15
ft	10	20	30	40	50	

X-RAY POWER

Chandra is 100 times more powerful than previous X-ray telescopes.

SPECIAL MIRRORS

The mirrors are coated in iridium and gold. These help reflect and focus the X-rays.

THE MOST
POWERFUL
X-RAY TELESCOPE
IN SPACE

ENERGY EFFICIENT

The foil-covered spacecraft houses the instruments needed to operate the X-ray telescope, while the solar panels supply the power. The entire observatory uses the same amount of electricity as a typical hairdryer.

GIANT EYE

JAMES WEBB SPACE TELESCOPE

Aiming to search for the first galaxies that formed after the Big Bang, the James Webb Space Telescope (JWST) will peer further into space than any other observatory before it. The telescope's immense primary mirror will gather in seven times more light than the Hubble Space Telescope, and will be powerful enough to detect infrared objects 400 times fainter than any current telescope can see. During its mission, the JWST will give scientists an unprecedented view of the Universe's history, allowing them to observe the formation of the earliest stars and planets, as well as investigating the potential for life.

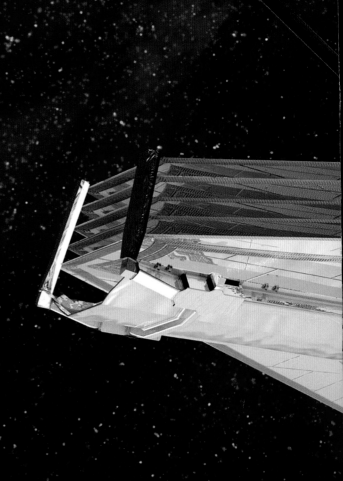

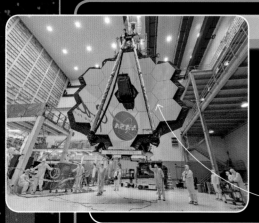

AT A GLANCE

- **TELESCOPE TYPE** Infrared space telescope

- **LAUNCHED BY** Ariane 5 ECA launch vehicle

- **LAUNCH DATE** Early 2020s

- **ORGANIZATIONS** NASA, ESA, and Canadian Space Agency (CSA)

  The telescope's primary mirror is made up of 18 hexagonal segments.

STATS AND FACTS

Named after James E Webb, a key figure in NASA's history, the JWST will travel to a point 1.5 million km (1 million miles) away from Earth – almost four times farther away than the Moon.

MIRROR DIAMETER

The telescope's primary mirror will be 6.5 m (21 ft) wide.

| m | 5 | 10 |
| ft | 16 | 32 |

The Hubble Space Telescope's main mirror is 2.4 m (8 ft) wide.

WEIGHT

The James Webb Space Telescope weighs 6,200 kg (14,000 lb).

GOLD LAYER

The primary mirror is coated with a layer of gold to improve the reflection of infrared light.

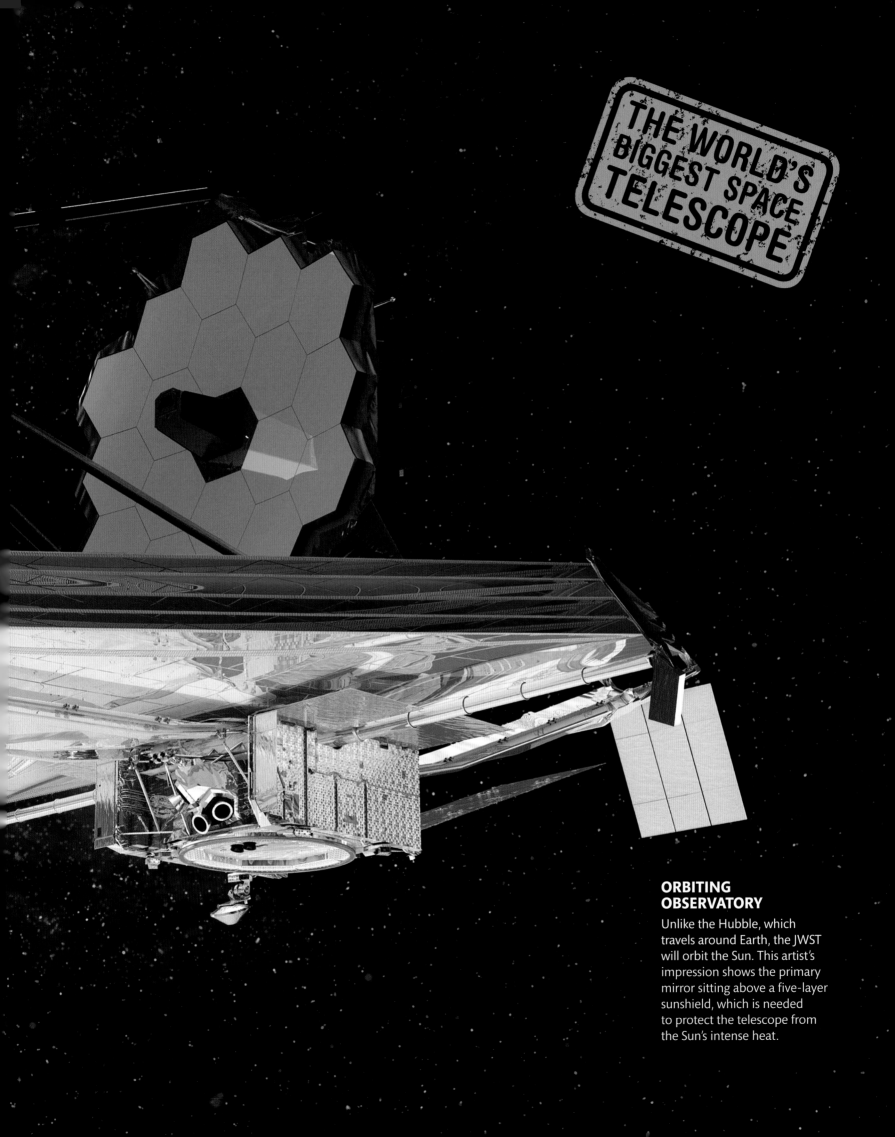

ORBITING OBSERVATORY

Unlike the Hubble, which travels around Earth, the JWST will orbit the Sun. This artist's impression shows the primary mirror sitting above a five-layer sunshield, which is needed to protect the telescope from the Sun's intense heat.

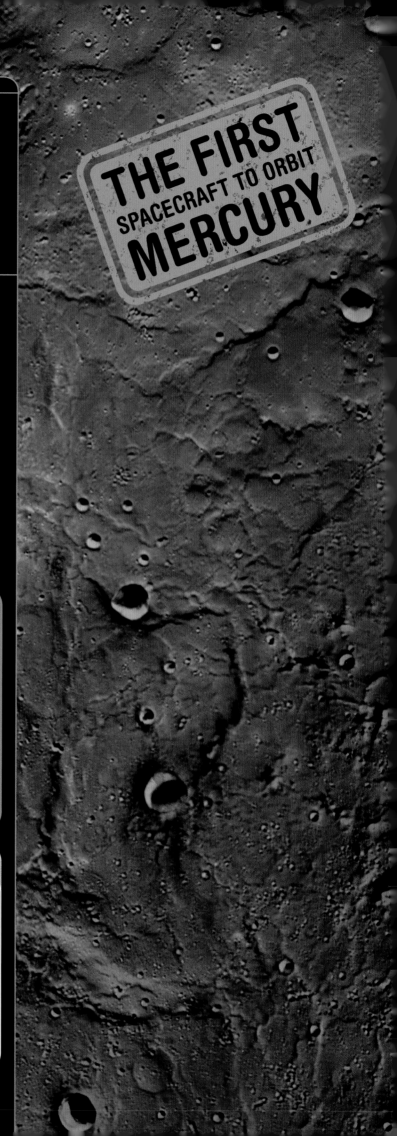

MAPPING MERCURY
MESSENGER

During the first mission to Mercury in 35 years, NASA's MESSENGER spacecraft mapped the planet in incredible detail. Launched from the USA in 2004, MESSENGER reached Mercury in 2011 and spent four years conducting observations. Instruments on the spacecraft photographed almost the entire surface, examined ice-filled craters at Mercury's north pole, and investigated its massive iron core. MESSENGER ran out of fuel in 2015 and was intentionally crashed into Mercury. This impact made a new mark on a surface already covered in craters from asteroids hitting the planet for billions of years.

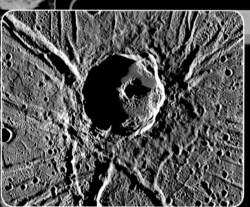

AT A GLANCE

- **SPACECRAFT TYPE** Orbiter
- **SIZE** 1.8 m x 1.42 m x 1.27 m (6 ft x 4 ft 8 in x 4 ft 2 in)
- **LAUNCH DATE** 3 August 2004
- **ORGANIZATION** NASA

Photographed by MESSENGER, the crater Apollodorus sits at the junction of many grooved valleys.

STATS AND FACTS

MESSENGER studied Mercury from orbit, sending the first photo of the rocky planet on 29 March 2011. During its mission, it took more than 200,000 photographs and mapped Mercury's entire surface.

DISTANCE

The craft travelled 7.9 billion km (4.9 billion miles) to reach Mercury in 2011.

ORBITS

MESSENGER completed a total of 4,100 orbits of Mercury.

PLANET SIZE

MESSENGER found that Mercury had shrunk by around 7 km (4 miles) since its formation.

SPEED

The craft's average speed was 147,625 km/h (91,730 mph).

HEAT SHIELD

This artist's impression shows MESSENGER in orbit around Mercury. Instruments were wrapped in protective foil, while a large, curved sunshade shielded the spacecraft against sizzling temperatures of 360°C (700°F) on the Sun-facing side.

SUN SEEKER

PARKER SOLAR PROBE

The Parker Solar Probe is set to break records. Not only is it the fastest spacecraft ever built, but it will also get seven times closer to the Sun than the previous record-holder, Helios-B. At its top speed, it would be fast enough to travel from London to New York City in less than 30 seconds. The probe will explore the outermost part of the Sun's atmosphere, the corona, to discover more about the solar wind. Its measurements will help predict changes in the space around our planet that could affect space technology and, ultimately, how we live on Earth.

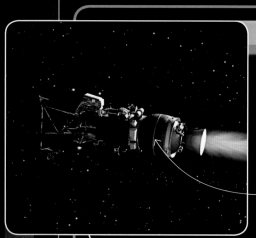

AT A GLANCE

- **SPACECRAFT TYPE** Solar orbiter
- **LAUNCH DATE** 12 August 2018
- **LAUNCH SITE** Cape Canaveral, USA
- **ORGANIZATION** NASA

A third rocket stage, jettisoned soon after launch, helped give the spacecraft the speed it needed to enter an orbit close to the Sun.

STATS AND FACTS

Between 2018 and 2025, the Parker Solar Probe will travel around the Sun 24 times, with each trip taking approximately 88 days. It will make its closest approach in 2025.

TEMPERATURE

The main instruments in the Parker Solar Probe will be kept cool at 29.5°C (85°F).

°C	500	1,000	1,500
°F	1,365		2,730

The temperature of the probe's heat shield will reach as high as 1,400°C (2,500°F).

TOP SPEED

The craft will have a top speed of 692,000 km/h (430,000 mph).

NAME

Parker is the first NASA probe named after a living person – the astrophysicist Eugene Parker.

FLYING INTO THE SUN

The Parker Solar Probe is protected by a 11.5 cm (4½ in) thick carbon heat shield known as the Thermal Protection System. This will keep the instruments behind it cool enough to run smoothly, while the shield itself bakes in the blazing heat of the Sun.

THE BIGGEST
PLANETARY ROVER

SELFIE ON MARS

Curiosity can even take selfies such as this one using a camera on the end of a long robotic arm. The rover also has a camera mounted on a mast that can look at rocks 7 m (3 ft) away, or take videos of the Martian landscape.

MARS EXPLORER
CURIOSITY

Part of NASA's Mars Science Laboratory mission, Curiosity is a car-sized robotic rover that set its aluminium wheels on the red planet in 2012. A trundling science laboratory, the rover carries a number of advanced science instruments, including its own weather station, cameras, lasers, and a precision drill to take small samples of Martian rock for analysis. It is studying the Martian climate and soil in the hunt for signs that there might be, or have been, life forms on the planet. In its first year on Mars, Curiosity sent back 36,700 images.

AT A GLANCE

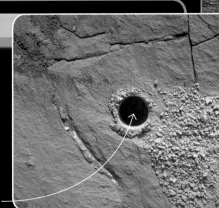

SPACE VEHICLE TYPE
Planetary rover

SIZE 3 m (10 ft) long

LAUNCH DATE 6 November 2011

LAUNCH SITE Cape Canaveral, Florida, USA

Drill hole dug by Curiosity

STATS AND FACTS

Curiosity's journey from Earth to Mars took nearly 9 months to complete. It had a complex landing that had never been attempted before, involving a parachute, rockets, and a sky crane, so that it could land gently on its wheels.

CAMERA EQUIPMENT

Curiosity is equipped with 17 cameras, some of which take true colour images.

SPEED

Curiosity moves at a relatively slow pace of 30 m (98 ft) per hour.

SOFT LANDING

The parachute that slowed Curiosity's descent was the largest ever built by NASA.

POWER SUPPLY

The rover started out with enough power to last at least 14 years.

INVESTIGATING ASTEROIDS

DAWN

Sent to explore Vesta and Ceres – the two most massive objects in the Asteroid Belt – NASA's Dawn spacecraft was the first to hop between two worlds. It began circling Vesta in 2011, making it the first spacecraft to orbit an asteroid. On Vesta, it found deep canyons and a mountain twice as tall as Mount Everest. It also confirmed that Vesta has a layered internal structure like Earth. In 2015, Dawn reached Ceres, becoming the first craft to visit a dwarf planet. It found salty mineral deposits on Ceres's icy surface and the possibility of a sub-surface ocean. Data on these ancient worlds provided scientists with insights into the early building blocks of the Solar System.

ENCIRCLING VESTA

Dawn orbits Vesta in this artist's impression. The spacecraft made 1,298 orbits of the asteroid, revealing a world with many craters.

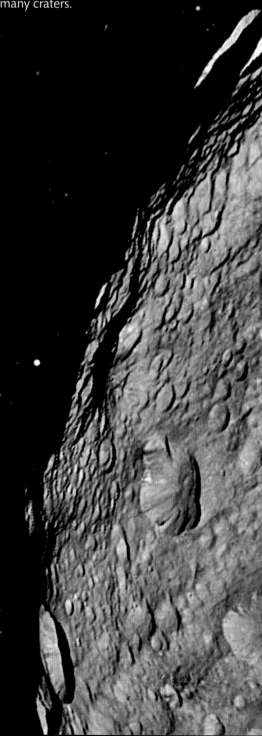

AT A GLANCE

- **SPACECRAFT TYPE** Orbiter
- **SIZE** 20 m (65 ft) wide, including the solar arrays
- **LAUNCH DATE** 27 September 2007
- **ORGANIZATION** NASA

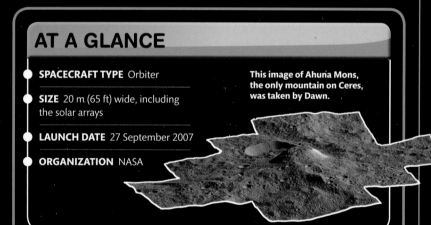

This image of Ahuna Mons, the only mountain on Ceres, was taken by Dawn.

STATS AND FACTS

Dawn travelled a distance of 2.8 billion km (1¾ billion miles) to reach Vesta, and then a further 1.5 billion km (930 million miles) to get to Ceres.

CRUISING SPACE

It took about 4 days for Dawn to accelerate from 0 to 97 km/h (0 to 60 mph).

AT VESTA

While orbiting Vesta, Dawn took about 31,000 photographs.

JOURNEY TO CERES

It took Dawn 2½ years to get from Vesta to Ceres, the largest object in the Asteroid Belt.

SPACE THRUSTERS

Dawn's ion thrusters used far less fuel than conventional rockets.

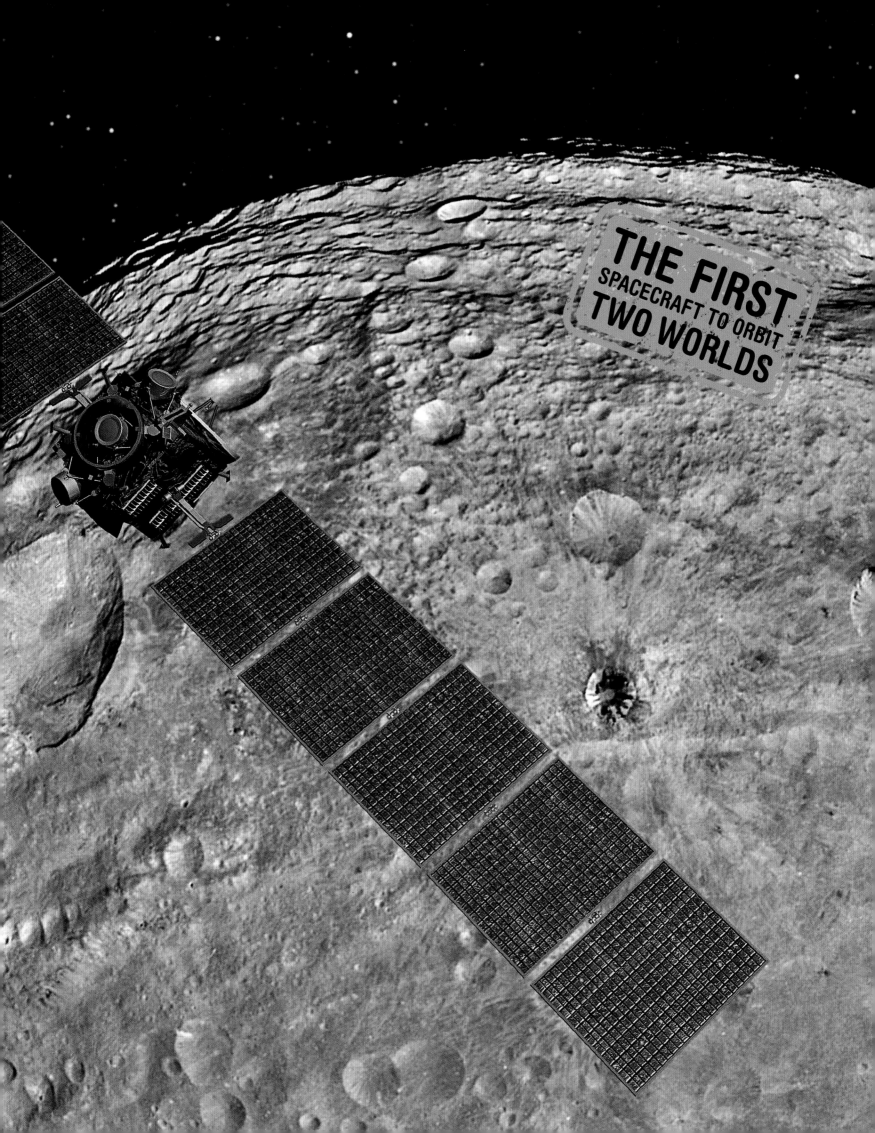

THE FIRST
SPACECRAFT TO ORBIT
TWO WORLDS

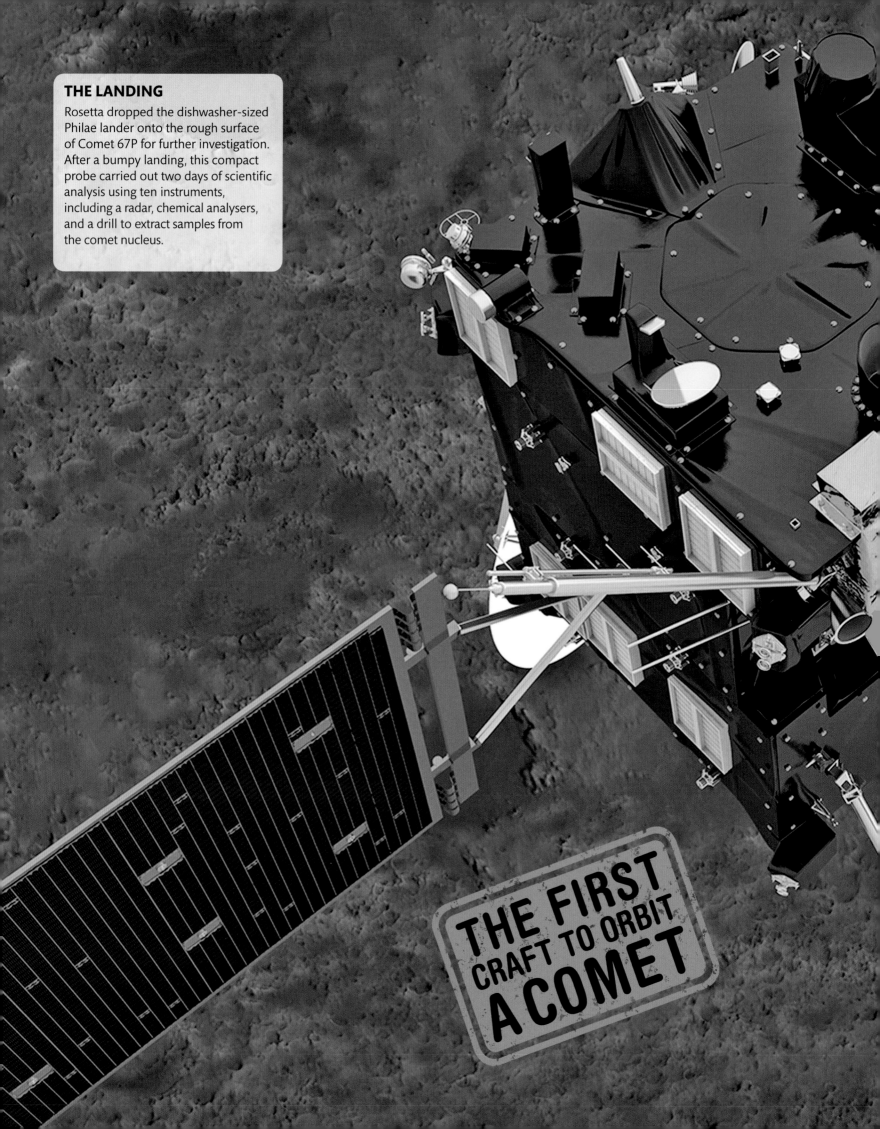

THE LANDING

Rosetta dropped the dishwasher-sized Philae lander onto the rough surface of Comet 67P for further investigation. After a bumpy landing, this compact probe carried out two days of scientific analysis using ten instruments, including a radar, chemical analysers, and a drill to extract samples from the comet nucleus.

THE FIRST CRAFT TO ORBIT A COMET

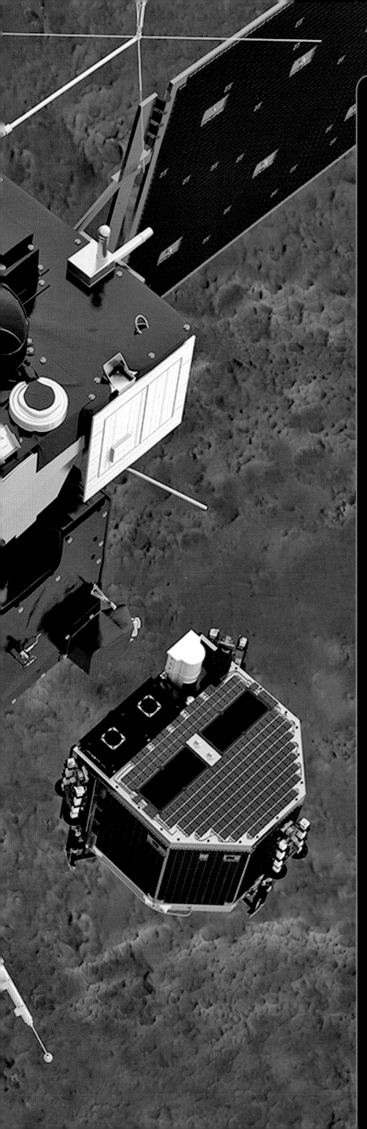

COMET CHASER

ROSETTA-PHILAE

Trillions of comets whizz through the most distant parts of the Solar System, but catching up with one is challenging. Rosetta did just that in 2014, becoming the first craft to orbit a comet, when it reached 67P after a ten-year journey of 6.4 billion km (4 billion miles). On-board instruments and cameras mapped the surface of the 4 km (2½ mile) wide icy blob on its path around the Sun. Rosetta also released its small Philae lander, which was the first spacecraft to land on a comet. Data from this mission are helping scientists learn about comet structure, which provides clues about the formation of our whole Solar System.

AT A GLANCE

- **SPACECRAFT TYPE** Orbiter (Rosetta) and lander probe (Philae)

- **LAUNCH DATE** 2 March 2004

- **MISSION END** 30 September 2016

- **ORGANIZATION** European Space Agency (ESA)

This spectacular image, taken by Rosetta, shows Comet 67P spewing out jets of gas and dust as the Sun warms it up.

STATS AND FACTS

At the start of the mission, Rosetta moved too slowly for its comet target, but picked up speed using the gravitational energy gathered from planets on the way to fly at more than 135,000 km/h (83,885 mph).

ASSISTED BY GRAVITY

To get enough speed, Rosetta made three flybys of Earth and one of Mars.

SOLAR POWER

Rosetta's solar panels gave it a wingspan longer than a basketball court.

DURATION

Rosetta-Philae's mission lasted for 12 years, 6 months, and 28 days.

ASTEROIDS

On its way to Comet 67P, Rosetta flew by two asteroids, gathering data on them.

EXPLORING JUPITER
JUNO

NASA'S Juno spacecraft entered the record books in 2016 when it became both the fastest and the most distant spacecraft powered by sunlight. Juno travelled 2.7 billion km (1¾ billion miles) from Earth to Jupiter and is on a mission to study how the gas giant formed, map its gravity and magnetic fields, and study its stormy atmosphere. With Jupiter so far from the Sun, Juno's solar panels and electronic systems are extremely efficient to compensate for limited sunlight. To get accurate measurements, the spacecraft is orbiting close to the gas giant. It also spins twice a minute, both for stability and to enable its on-board instruments to take a detailed look at the planet.

AT A GLANCE

- **SPACECRAFT TYPE** Orbiter
- **LAUNCH DATE** 5 August 2011
- **MISSION ENDING** 2021
- **ORBITAL PERIOD** Every 53 days around Jupiter

This image from Juno shows Jupiter's south pole and its surprising blue-green colour.

STATS AND FACTS

Juno is the first spacecraft exploring the outer planets to be powered by solar panels. All others have relied on radioactive power sources. The spacecraft is equipped with nine scientific instruments for gathering data.

SPEED

Juno entered orbit around Jupiter at 210,000 km/h (130,488 mph).

SIZE

The spacecraft is about the size of a basketball court.

RINGS

Juno took the first detailed image of Jupiter's rings from within the ring system.

JUNO'S NAME

In Roman mythology, Juno was Jupiter's wife, Saturn's daughter and Mars's mother.

THE MOST DISTANT SOLAR-POWERED SPACECRAFT

SOLAR POWER

Juno's three wing-like solar panels together produce about 500 watts of power, which is sufficient to fly the spacecraft and run its instruments. On the end of one solar panel is the magnetometer, which measures the magnetic field.

THE FIRST CRAFT TO ORBIT SATURN

LANDMARK MISSION
CASSINI-HUYGENS

In 2004, the spacecraft Cassini travelled through a gap between Saturn's rings to become the first ever craft to orbit the planet. The size of a minibus, it was the largest spacecraft ever built. While studying Saturn, Cassini circled the gas giant 294 times, advancing our knowledge of this ringed world. It also discovered six new moons orbiting Saturn and lakes of liquid methane and ethane on Titan, Saturn's largest moon. In December 2004, the Huygens probe separated from the Cassini orbiter and landed on Titan, where it found evidence of Earth-like features. With its fuel finished, Cassini's mission came to an end in 2017, when it ploughed through Saturn's atmosphere.

AT A GLANCE

- **SPACECRAFT TYPE** Orbiter (Cassini) and lander (Huygens)
- **SIZE OF SPACECRAFT** 6.8 m (22 ft)
- **LAUNCH DATE** 15 October 1997
- **MISSION DURATION** 19 years and 335 days

This artist's impression shows the Huygens probe parachuting onto the surface of Titan.

STATS AND FACTS

The spacecraft was named after Italian astronomer Giovanni Cassini. In the 1700s, he discovered Saturn's rings as well as four of its moons.

FLY-BYS

Cassini made a total of 162 fly-bys past Saturn's moons.

DATA

The mission produced about 635 GB of valuable science data.

DISTANCE TRAVELLED

Cassini travelled a total of 7.8 billion km (4.9 billion miles).

IMAGES

Cassini took 453,048 images during its mission.

FINAL MISSION
Before the end of its mission, Cassini began a series of 22 dives to provide scientists with spectacular close-up views of Saturn's rings.

DISCOVERING NEPTUNE

In this artist's illustration, Voyager 2 is approaching our Solar System's most remote planet, Neptune. Various instruments onboard Voyager 2 helped scientists to discover Neptune's rings, moons, and storms, which are the strongest in the Solar System.

MARATHON MISSION
VOYAGER 2

The twin spacecraft Voyager 1 and 2 have flown further into space than any other craft to date. NASA launched these spacecraft in 1977 on an epic journey to study the most distant planets of our Solar System. Although Voyager 1 has travelled the greatest distance, Voyager 2 is the only craft to study all four giant planets at close range. It passed Jupiter in 1979, Saturn in 1981, Uranus in 1986, and finally Neptune in 1989 after travelling more than 7 billion km (4 billion miles). Voyager 2 has taken detailed photographs of these planets and their moons. It has now ventured into interstellar space, providing information on an unexplored part of space.

AT A GLANCE

- **SPACECRAFT TYPE** Flyby

- **LAUNCH DATE** 20 August 1977

- **LAUNCH SITE** Cape Canaveral, Florida, USA

- **MISSION DURATION** Originally planned to last 5 years, the mission is still ongoing

Both missions carry a Voyager Golden Record, which is a disc containing a message from Earth.

STATS AND FACTS

The two Voyager Golden Records feature images, sounds, music, and greetings in 55 different languages. Graphics on the cover show any alien life forms how to play the record.

SENDING DATA

Voyager 2 is so far away that it takes over 16 hours for signals from the spacecraft to reach Earth.

hours | 12 | 24

It takes Curiosity up to about 20 minutes to send signals from Mars to Earth.

STORM

Voyager 2 discovered Neptune's Great Dark Spot in 1989, which was a colossal storm the size of Earth.

GEYSERS

Voyager 2 discovered geysers on Neptune's moon Triton.

THE FIRST
SPACECRAFT
TO VISIT PLUTO

LONG-DISTANCE EXPLORER
NEW HORIZONS

As part of NASA's New Frontiers programme, New Horizons left Earth in 2006 on a mission to become the first visitor to the dwarf planet Pluto. During the epic journey, the speedy spacecraft rocketed past the Moon in nine hours and passed Jupiter a year later. But despite travelling 60 times faster than a jet airliner, it took almost a decade to reach the outer edges of our Solar System. In 2015, New Horizons finally flew within 12,500 km (7,750 miles) of Pluto, revealing its spectacular surface of icy mountains, volcanoes, canyons, and glaciers. After visiting Pluto, it journeyed deeper into space and, on 1 January 2019, flew past the Kuiper Belt object 2014 MU$_{69}$ (nicknamed Ultima Thule) – the most remote object targeted by a spacecraft.

AT A GLANCE

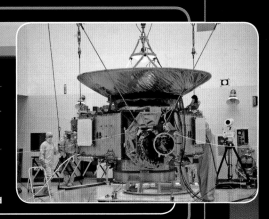

- **SPACECRAFT TYPE** Flyby
- **LAUNCH DATE** 19 January 2006
- **ARRIVAL** At Pluto: 14 July 2015; at 2014 MU$_{69}$: 1 January 2019
- **ORGANIZATION** NASA

New Horizons being assembled

APPROACHING PLUTO

New Horizons is seen flying past Pluto with its biggest moon Charon in the distance in this artist's impression. Scientific instruments and cameras onboard took photographs and recorded data, which was transmitted back to Earth via the craft's large dish antenna.

STATS AND FACTS

When NASA launched New Horizons, Pluto was considered a planet in the Solar System. By the time the spacecraft arrived, Pluto had been reclassified as a dwarf planet.

MICROCHIP

The microchip on New Horizons is also found in the PlayStation games consoles.

DISTANCE

Since passing Pluto, New Horizons crossed 1.1 million km (683,500 miles) of space each day.

CLOSEST APPROACH

At its closest, New Horizons got to within 12,550 km (7,800 miles) of Pluto's surface.

CONTACT

Dish antennae in the USA, Spain, and Australia are used to keep in touch with the spacecraft.

ASTEROID EXPLORER
HAYABUSA2

In 2018, a Japanese spacecraft carrying four probes and a sample return capsule arrived at a near-Earth asteroid – 162173 Ryugu – after travelling a distance of 3.2 billion km (2 billion miles). Over the next 18 months, the craft deployed its probes – three tiny rovers and a lander – to examine the asteroid from all angles and blast holes in its surface to study the interior rock. Samples of the asteroid will return to Earth inside a capsule, due to arrive in December 2020. By this time the spacecraft will have travelled more than 5 billion km (3 billion miles), the same as 35 journeys from Earth to the Sun.

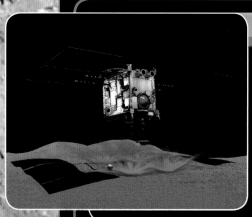

AT A GLANCE

- **SPACECRAFT TYPE** Orbiter, lander, rover, sample return
- **LAUNCH DATE** 3 December 2014
- **LAUNCH SITE** Tanegashima Space Center, Japan
- **ORGANIZATION** Japan Aerospace Exploration Agency (JAXA)

An artist's impression of Hayabusa2 hovering over asteroid Ryugu's surface.

STATS AND FACTS

Hayabusa2 follows the Hayabusa mission of 2010, which faced technical problems but successfully returned samples of asteroid Itokawa to Earth.

SOLAR POWER

The craft's twin solar arrays are 4.2 m (13.8 ft) long.

PROBES

The probes carried by Hayabusa2 are among the lightest ever deployed in the Solar System.

EXPLOSIVE PROJECTILE

In 2019, a projectile was fired at the asteroid using an explosive charge to excavate a hole.

BATTERY LIFE

The lander probe had just about 16 hours of battery to complete its operation.

LAUNCH WEIGHT

609 KG
(1,343 LB)

ON TARGET

Hayabusa2 is seen here closing in on asteroid Ryugu. Measuring 850 m (2,800 ft) wide, this carbon-rich asteroid features large craters and boulders on its surface, as well as a ring of peaks around its equator.

THE FIRST MULTI-PROBE INVESTIGATION OF AN ASTEROID

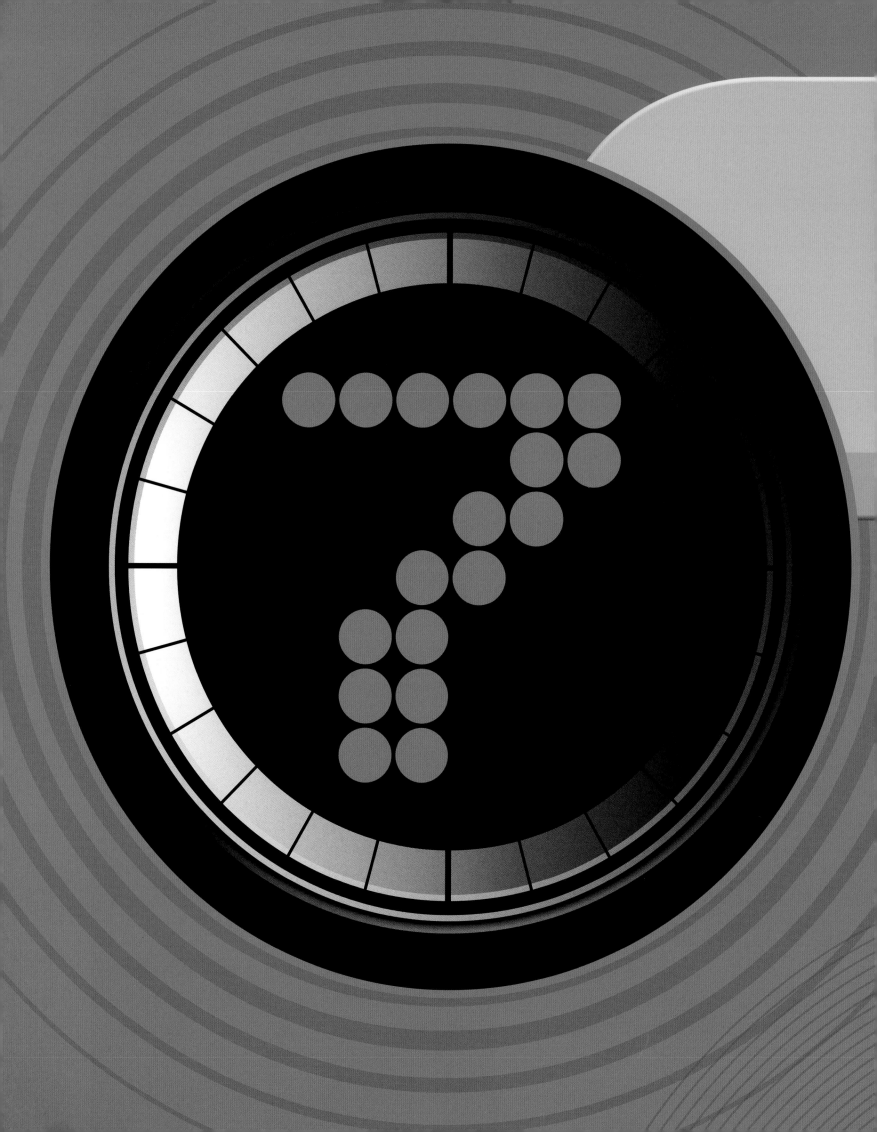

PEOPLE IN SPACE

Since 1961, more than
500 people have travelled
into space from Earth. Blasting
off on rocket-powered launch
vehicles, these intrepid explorers
have orbited Earth, inhabited
space stations, and even set foot
on the Moon. Their experiences
have pushed the boundaries
of human ingenuity
and endurance.

FIRST CREWED SPACECRAFT
VOSTOK 1

History was made in 1961 when a person travelled from planet Earth into space for the first time. Launched by a Soviet rocket, the tiny Vostok 1 spacecraft carried 27-year-old cosmonaut Yuri Gagarin. After completing one orbit of Earth, Gagarin made a safe return, landing by parachute in a field near a Soviet town. He proved that people could survive space travel. Five further Vostok spacecraft followed this landmark mission.

AT A GLANCE

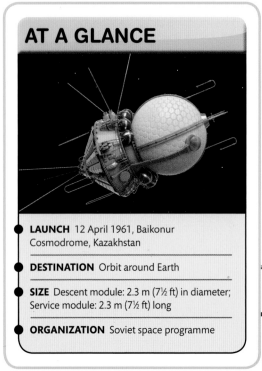

- **LAUNCH** 12 April 1961, Baikonur Cosmodrome, Kazakhstan

- **DESTINATION** Orbit around Earth

- **SIZE** Descent module: 2.3 m (7½ ft) in diameter; Service module: 2.3 m (7½ ft) long

- **ORGANIZATION** Soviet space programme

The service module housed the electrical power supply, the engine, and a small rocket to turn the craft around.

The retrorocket engine fired for 266 seconds to blast the spacecraft back into Earth's atmosphere.

Communications antennae

A series of tanks containing nitrogen and oxygen were mixed together to provide breathable conditions on board.

STATS AND FACTS

327 KM (203 MILES)
MAXIMUM HEIGHT REACHED

Scientific curiosity and the desire to beat the USA into space saw the Vostok programme begin testing in 1960.

MISSION DURATION

Vostok 1 was the shortest mission of the Vostok programme, lasting for 1 hour and 48 minutes.

| hours | 60 | 120 |

Vostok 5 was the longest mission, lasting for 119 hours.

ROCKET LAUNCH

Vostok 1 was carried into space on board a converted ballistic missile.

SPACESUIT

The spacesuit featured a mirrored sleeve to view all instruments.

How it worked

Vostok 1 consisted of a service module and a descent module. The uncrewed cone-shaped service module carried the engine, air tanks, and fuel, while the ball-shaped descent module carried Gagarin. He had no control over the spacecraft – the entire mission was remotely controlled back at the space centre. Gagarin was given instructions on how to override the control systems in an emergency. After Vostok's orbit, metal straps holding the two modules together were blown away, so the descent module could separate and return to Earth.

The walls were heavily padded to help reduce the noise and vibrations.

The instrument panel displayed the on-board temperature, air pressure, and position of the spacecraft.

INSIDE THE DESCENT MODULE

Apart from the instrument panel, the bulky ejector seat, and Gagarin himself, there was no room for anything else in the cramped descent module. Gagarin lay strapped to his seat throughout the mission. A small window provided a view of space.

These metal straps connected the two modules.

The heat shield protected the descent module from temperatures of 3,000°C (5,400°F) on re-entering the atmosphere.

This entry hatch was blown off before Gagarin's ejector seat fired him out of the capsule.

BACK TO EARTH

About 7 km (4.5 miles) above Earth, the hatch on the descent module was blown open by explosives. The ejection seat fired, carrying the astronaut clear. Parachutes brought both the module and the astronaut safely down to Earth.

These radio antennae allowed Gagarin to communicate with ground control.

FIRST MAN IN SPACE

Yuri Gagarin was a Soviet air force pilot before training as a cosmonaut. He was chosen from more than 3,000 candidates to fly Vostok 1.

FIRST WOMAN IN SPACE

In 1963, Soviet skydiver Valentina Tereshkova became the first woman in space. She orbited Earth 48 times during her three-day mission onboard Vostok 6.

SPACE WORKHORSE
SOYUZ

For more than 50 years and 140 missions, the three-module spacecraft called Soyuz has been the backbone of the Russian human spaceflight programme. Soyuz means "union" in Russian, an apt name given that the craft regularly docks with others in space. Since the retirement of NASA's space shuttle in 2011, it is the only vessel capable of ferrying astronauts to and from the International Space Station.

The service module houses the control systems, electric power supply, and communication systems – everything needed to support its mission.

Digital Soyuz

Soyuz MS, the latest member of the Soyuz family, first flew in 2016. It is 7.48 m (24½ ft) in length, and features upgraded electronics and flight computers as well as more efficient solar panels and an additional battery.

The solar panels generate all electrical power and can be rotated to face towards the Sun.

AT A GLANCE

Soyuz MS-04 launches on a Soyuz-FG rocket.

FIRST SOYUZ MISSION WITH A CREW
23 April 1967

CREW Three: a commander, a co-pilot, and a flight engineer

ORGANIZATION Soviet space programme/ Roscosmos (Russian Federal Space Agency)

NOTABLE ACHIEVEMENTS Ferrying astronauts to Salyut, Mir, and ISS; orbiting the Moon in 1968; and docking with an Apollo module in 1975

This is one of the radar antennae that help to judge distances to objects when the craft is docking in space.

STATS AND FACTS

Four major generations and multiple versions of the Soyuz spacecraft have been built. It is the world's safest, cheapest spacecraft for ferrying astronauts.

LIVING SPACE

Soyuz's orbital module has only 5 cubic metres (177 cubic ft) of habitable space.

cubic metre	20	40	60	80	100

cubic ft	1,800	3,600

NASA's Space Shuttle had 65.8 cubic metres (2,325 cubic ft) of habitable space.

EVER-PRESENT

Since 2000, at least one Soyuz has always been docked with the ISS.

MOON VISITORS

In 1968, a Soyuz flew the first creatures (two tortoises, flies, and mealworms) around the Moon.

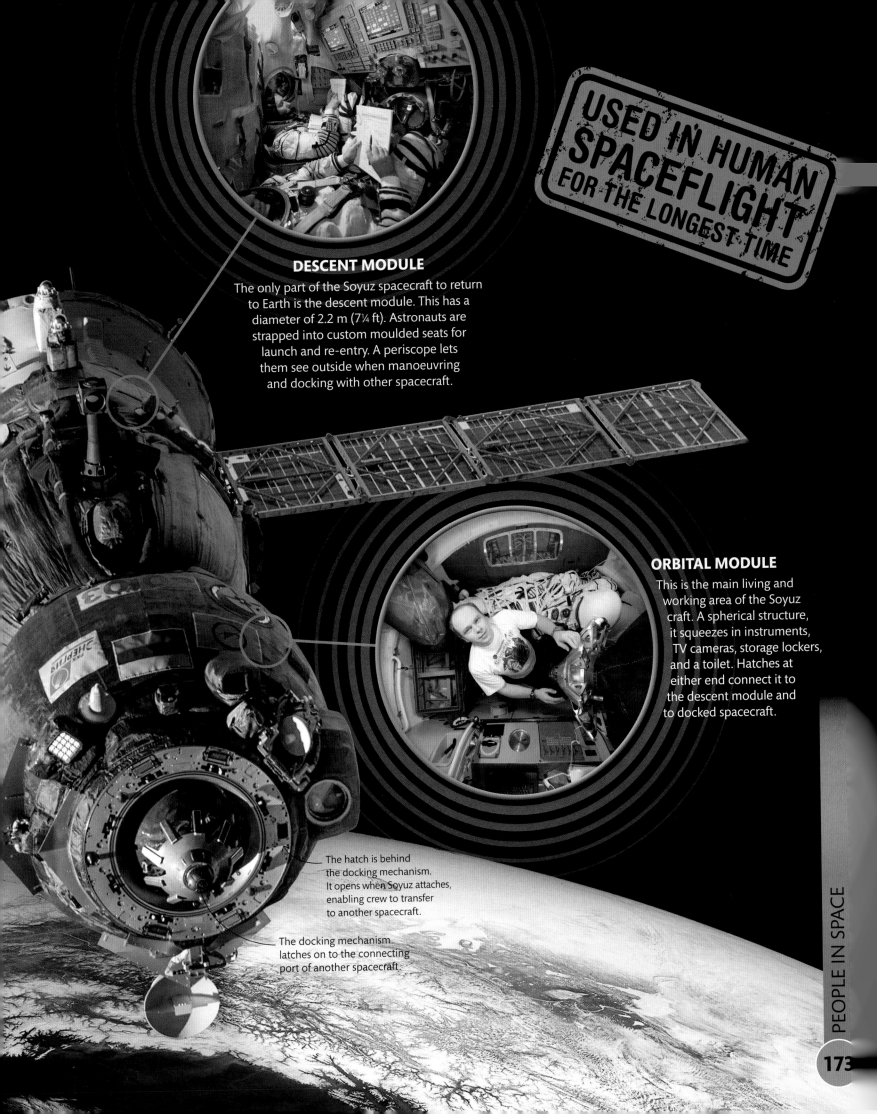

DESCENT MODULE

The only part of the Soyuz spacecraft to return to Earth is the descent module. This has a diameter of 2.2 m (7¼ ft). Astronauts are strapped into custom moulded seats for launch and re-entry. A periscope lets them see outside when manoeuvring and docking with other spacecraft.

ORBITAL MODULE

This is the main living and working area of the Soyuz craft. A spherical structure, it squeezes in instruments, TV cameras, storage lockers, and a toilet. Hatches at either end connect it to the descent module and to docked spacecraft.

The hatch is behind the docking mechanism. It opens when Soyuz attaches, enabling crew to transfer to another spacecraft.

The docking mechanism latches on to the connecting port of another spacecraft.

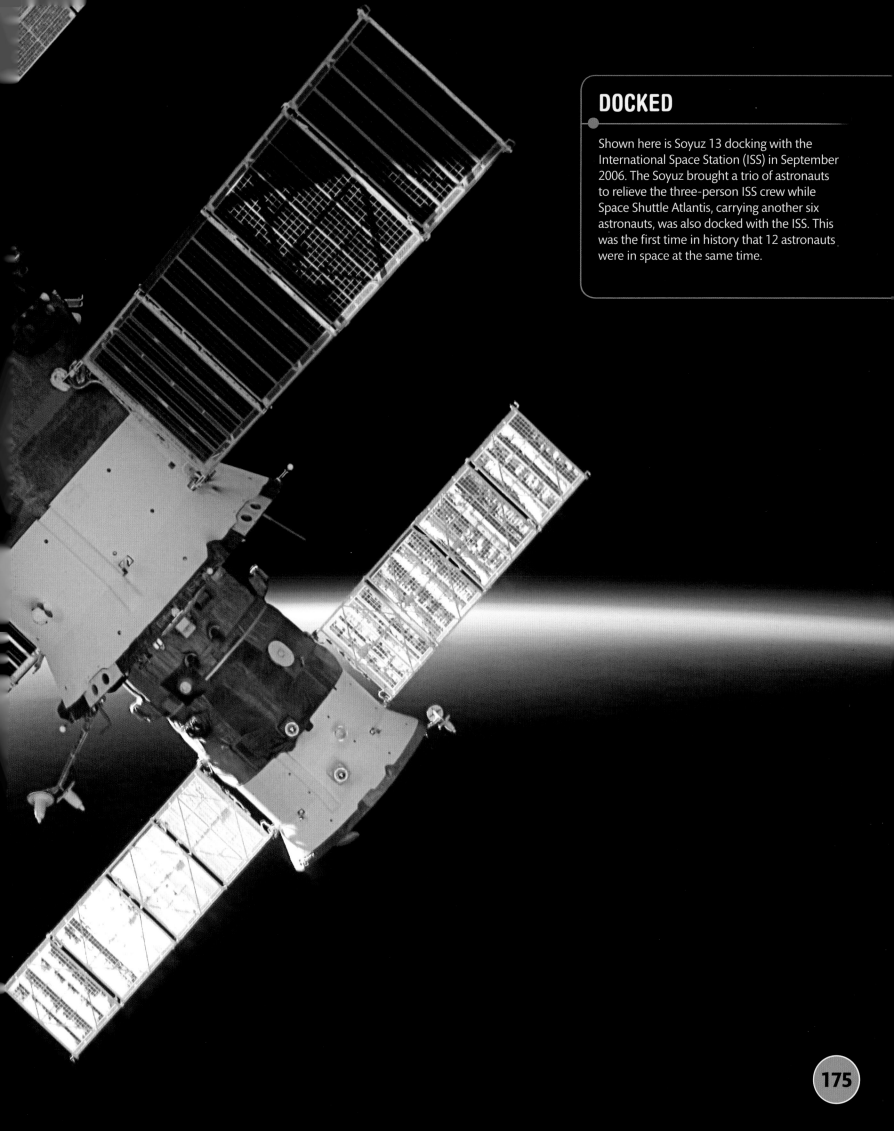

DOCKED

Shown here is Soyuz 13 docking with the International Space Station (ISS) in September 2006. The Soyuz brought a trio of astronauts to relieve the three-person ISS crew while Space Shuttle Atlantis, carrying another six astronauts, was also docked with the ISS. This was the first time in history that 12 astronauts were in space at the same time.

ROBUST ROCKET
SATURN V

The tallest, heaviest, and most powerful rocket ever flown is Saturn V. At lift-off this NASA launch vehicle, which was the same height as a 36-storey building, weighed more than 35 space shuttles. In the 1960s and the 1970s, Saturn V rockets launched Apollo spacecraft, including the one that put the first people on the Moon in 1969. Each Saturn V consisted of three separate rocket stages, each with its own engine and fuel. After each stage used up its fuel, it was discarded to reduce weight. The first stage took the spacecraft to an altitude of 60 km (37 miles) and a speed of almost 10,000 km/h (6,000 mph). By the third and final stage the spacecraft had soared to 190 km (118 miles) high at 25,000 km/h (15,500 mph).

AT A GLANCE

- **VEHICLE TYPE** Three-stage launch vehicle
- **FIRST LAUNCH** 9 November 1967, Apollo 4 uncrewed test mission
- **LAST LAUNCH** 14 May 1973, carrying Skylab space station into orbit
- **ORGANIZATION** NASA

Apollo 11 – the first mission to put people on the Moon – launched in July 1969.

THE MOST POWERFUL LAUNCH VEHICLE EVER FLOWN

STATS AND FACTS

Saturn V went from initial plans to first launch in just six years with the input of more than 20,000 companies. A total of 15 Saturn V rockets were built.

HEIGHT

At 111 m (364 ft) high, Saturn V was taller than the Statue of Liberty.

m	60	120
ft	197	394

WEIGHT

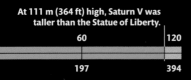

The Saturn V rocket had a total weight of 2,700 tonnes when fully fuelled.

tonnes		
1,000	2,000	3,000

The Soviet launch vehicle Proton had a weight of 683 tonnes when fuelled.

TOTAL LAUNCHES
13

POWERFUL ENGINES

The first stage of a Saturn V rocket is seen here on display at NASA's Johnson Space Center. The five first-stage rocket engines had the power of 30 jumbo jets. They burned 1,789 kg (3,945 lb) of liquid oxygen and 788 kg (1,738 lb) of fuel every second and fired for two minutes, 30 seconds.

LUNAR LANDING

APOLLO 11

THREE MODULES

The Apollo spacecraft featured three modules. The two-person Lunar Module travelled to the Moon's surface and could split into two separate stages. Meanwhile, astronaut Michael Collins piloted the Command and Service modules as they orbited 100 km (62 miles) above, providing a communications link to Earth.

COMMAND MODULE
The astronauts began and ended their mission in this cone-shaped capsule, the only part of the Apollo spacecraft designed to return to Earth. After re-entering Earth's atmosphere, it parachuted into the ocean.

SERVICE MODULE
Connected to the Command Module for most of the mission, this module supplied oxygen, water, and electrical power to the spacecraft and had a powerful engine. The Service Module separated shortly before re-entry and burned up in Earth's atmosphere.

LUNAR MODULE ASCENT STAGE
Astronauts stood up inside this stage's cabin when flying the Lunar Module. On leaving the Moon, it separated from the Descent Stage, which it used as a launch pad.

LUNAR MODULE DESCENT STAGE
Three quarters of this module's weight was fuel for a large rocket engine, which controlled the spacecraft's descent to the lunar surface.

American astronaut Neil Armstrong made history on 21 July 1969 when he became the first person to walk on the Moon, after successfully landing the Apollo 11 Lunar Module with only seconds of fuel remaining. This was the climax of an eight-year space race between the USA and the Soviet Union – the world's two superpowers – to put people on the Moon. A total of 12 astronauts set foot on the Moon during NASA's Apollo program.

FIRST PEOPLE ON THE MOON

Neil Armstrong stepped onto the Moon first, followed 19 minutes later by Edwin Buzz Aldrin, seen here. During more than 21 hours on the Moon, the pair planted the American flag, collected samples, took photographs, and slept onboard the Lunar Module.

STATS AND FACTS

The Apollo 11 Lunar Module started its journey with 8,200 kg (18,000 lb) of fuel in its descent-stage tanks but landed on the Moon with less than one minute's worth of fuel left.

DURATION

The Apollo 11 mission to the Moon lasted 8 days, 3 hours, and 18.6 minutes.

COMPUTING POWER
Apollo 11's computer was less powerful than a modern mobile phone.

THRUSTERS

The Lunar Module was steered through space by 16 small jet thrusters, arranged in four groups.

QUICK FIX
A broken engine switch was replaced by a metal pen before the Lunar Module Ascent Stage took off from the Moon.

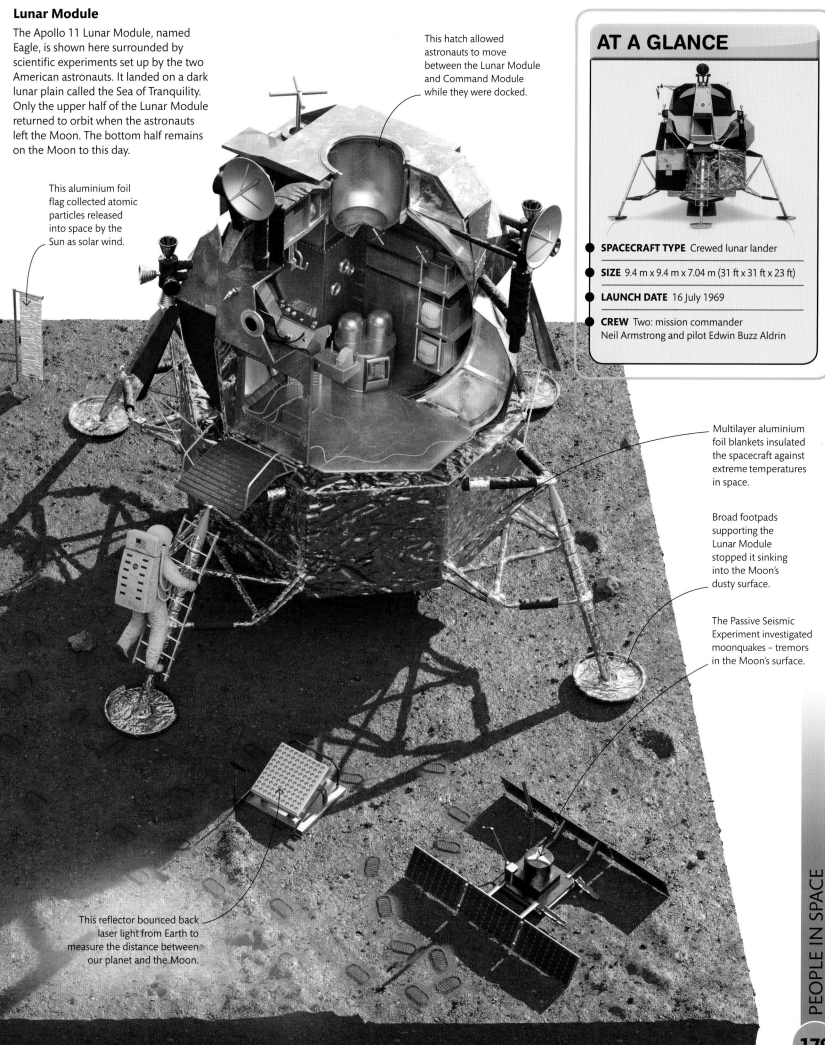

Lunar Module

The Apollo 11 Lunar Module, named Eagle, is shown here surrounded by scientific experiments set up by the two American astronauts. It landed on a dark lunar plain called the Sea of Tranquility. Only the upper half of the Lunar Module returned to orbit when the astronauts left the Moon. The bottom half remains on the Moon to this day.

This hatch allowed astronauts to move between the Lunar Module and Command Module while they were docked.

This aluminium foil flag collected atomic particles released into space by the Sun as solar wind.

Multilayer aluminium foil blankets insulated the spacecraft against extreme temperatures in space.

Broad footpads supporting the Lunar Module stopped it sinking into the Moon's dusty surface.

The Passive Seismic Experiment investigated moonquakes – tremors in the Moon's surface.

This reflector bounced back laser light from Earth to measure the distance between our planet and the Moon.

AT A GLANCE

- **SPACECRAFT TYPE** Crewed lunar lander
- **SIZE** 9.4 m x 9.4 m x 7.04 m (31 ft x 31 ft x 23 ft)
- **LAUNCH DATE** 16 July 1969
- **CREW** Two: mission commander Neil Armstrong and pilot Edwin Buzz Aldrin

MOONWALK

NASA astronauts Neil Armstrong and Edwin Buzz Aldrin made history when they became the first people to walk on the Moon. During their mission, they spent more than two hours on the lunar surface, deploying experiments and collecting rocks. In this photo taken by Armstrong (who casts his shadow), Aldrin stands in front of Apollo 11's Lunar Module.

SPACE CARGO
This photograph was taken from the International Space Station in July 2006. It shows Discovery at the start of its 12-day mission, preparing to dock with the station. The open doors reveal the cargo cylinder holding supplies and equipment.

REUSABLE SHUTTLE
SPACE SHUTTLE DISCOVERY

Record-breaking Discovery clocked up 39 missions in total, more than any other space shuttle in NASA's reusable fleet. Launched vertically using giant booster rockets, the craft orbited Earth at speeds topping 28,000 km/h (17,400 mph) for up to two weeks before gliding back home and landing like an aircraft on a runway. During 365 days in space, Discovery covered a distance equal to 620 Earth-to-Moon trips. Missions included launching the Hubble Space Telescope, visiting the Mir space station, and helping to build the International Space Station.

AT A GLANCE

SPACECRAFT TYPE Orbiter

FIRST FLIGHT 30 August 1984

FINAL FLIGHT 24 February 2011

ORGANIZATION NASA

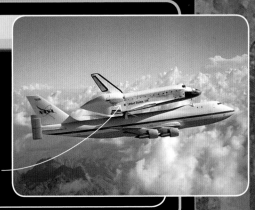

Discovery hitches a ride on NASA's modified carrier airplane on its way to the Kennedy Space Center, Florida, USA.

STATS AND FACTS

Discovery was one of NASA's five space shuttles; the other four were Columbia, Challenger, Atlantis, and Endeavour. On a typical mission, the spacecraft carried up to 7 crew members and each mission lasted from 5 to 16 days.

FEMALE PILOT

Discovery was the first American spacecraft with a female pilot, Eileen Collins, in 1995.

WEIGHT OF PAYLOAD BAY

The size of a school bus, the payload bay carried up to 25 tonnes of cargo.

ORBITS

The space shuttle made 5,830 orbits of Earth during 27 years of service.

PROTECTIVE TILES

More than 23,000 tiles protected Discovery against atmospheric heat on its return to Earth.

MOST FLIGHTS MADE BY A SPACE SHUTTLE

MINI SPACE SHUTTLE

This small, reusable spacecraft is called the Dream Chaser. It will deliver cargo to the International Space Station from late 2020. The craft is about a quarter of the length of a space shuttle, and its key feature – the angled wings – are designed to fold up when it is launched on NASA's Atlas V rocket. Following its return from space, the wings will open out, enabling the spacecraft to glide back to Earth.

LARGEST SPACE HABITAT
INTERNATIONAL SPACE STATION

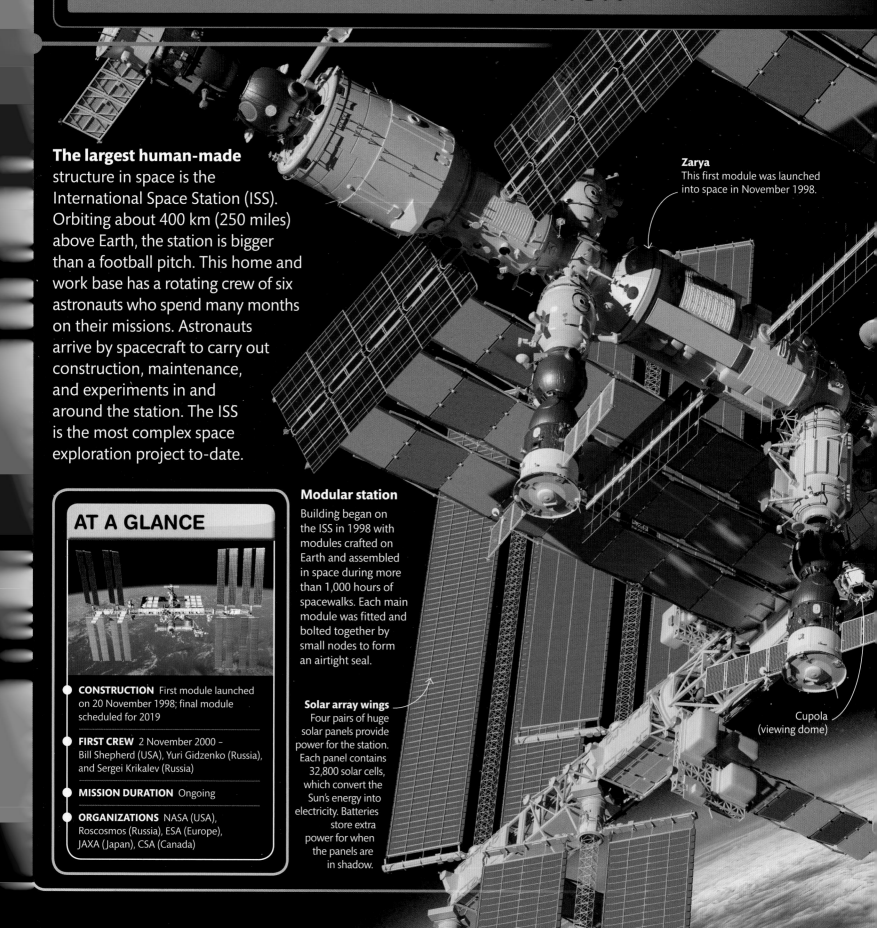

The largest human-made structure in space is the International Space Station (ISS). Orbiting about 400 km (250 miles) above Earth, the station is bigger than a football pitch. This home and work base has a rotating crew of six astronauts who spend many months on their missions. Astronauts arrive by spacecraft to carry out construction, maintenance, and experiments in and around the station. The ISS is the most complex space exploration project to-date.

Zarya
This first module was launched into space in November 1998.

Modular station
Building began on the ISS in 1998 with modules crafted on Earth and assembled in space during more than 1,000 hours of spacewalks. Each main module was fitted and bolted together by small nodes to form an airtight seal.

Solar array wings
Four pairs of huge solar panels provide power for the station. Each panel contains 32,800 solar cells, which convert the Sun's energy into electricity. Batteries store extra power for when the panels are in shadow.

Cupola (viewing dome)

AT A GLANCE

- **CONSTRUCTION** First module launched on 20 November 1998; final module scheduled for 2019

- **FIRST CREW** 2 November 2000 – Bill Shepherd (USA), Yuri Gidzenko (Russia), and Sergei Krikalev (Russia)

- **MISSION DURATION** Ongoing

- **ORGANIZATIONS** NASA (USA), Roscosmos (Russia), ESA (Europe), JAXA (Japan), CSA (Canada)

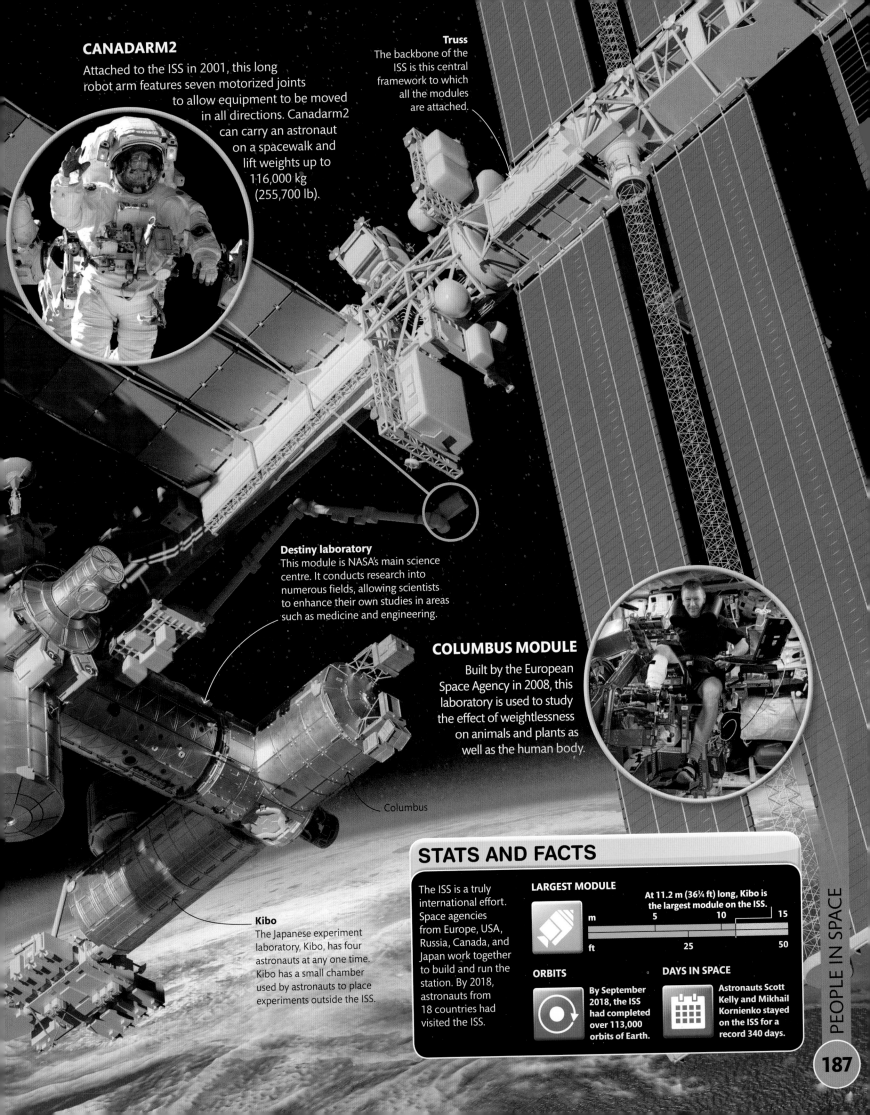

CANADARM2

Attached to the ISS in 2001, this long robot arm features seven motorized joints to allow equipment to be moved in all directions. Canadarm2 can carry an astronaut on a spacewalk and lift weights up to 116,000 kg (255,700 lb).

Truss
The backbone of the ISS is this central framework to which all the modules are attached.

Destiny laboratory
This module is NASA's main science centre. It conducts research into numerous fields, allowing scientists to enhance their own studies in areas such as medicine and engineering.

COLUMBUS MODULE

Built by the European Space Agency in 2008, this laboratory is used to study the effect of weightlessness on animals and plants as well as the human body.

Columbus

Kibo
The Japanese experiment laboratory, Kibo, has four astronauts at any one time. Kibo has a small chamber used by astronauts to place experiments outside the ISS.

STATS AND FACTS

The ISS is a truly international effort. Space agencies from Europe, USA, Russia, Canada, and Japan work together to build and run the station. By 2018, astronauts from 18 countries had visited the ISS.

LARGEST MODULE

At 11.2 m (36¾ ft) long, Kibo is the largest module on the ISS.

| m | 5 | 10 | 15 |
| ft | 25 | | 50 |

ORBITS
By September 2018, the ISS had completed over 113,000 orbits of Earth.

DAYS IN SPACE
Astronauts Scott Kelly and Mikhail Kornienko stayed on the ISS for a record 340 days.

EXPERIMENTS IN SPACE

Astronauts conduct experiments on board the International Space Station to find out how extended periods living in microgravity affect the human body. NASA astronaut Karen Nyberg is seen here inside the Destiny laboratory, performing an eye exam on herself. The data collected from such experiments is analysed to find out how vision may be affected by living in space.

SURVIVAL SUIT
EMU SPACESUIT

A spacesuit provides protection against the many hazards of space. This robust multi-layered clothing ensures astronauts can survive the lack of air, extreme temperatures, solar radiation, and dangerous flying debris. NASA's EMU spacesuit comes equipped with advanced technological features for astronauts making spacewalks from the International Space Station. Supplies of oxygen, water, and power contained inside the pressurized suit keep astronauts safe in space.

AT A GLANCE

- **FIRST USE OF EMU SPACESUIT** 1998 (original version of EMU was introduced in 1981)

- **TYPE OF SUIT** Pressurized spacesuit worn to perform extravehicular activities (EVAs) outside spacecraft

- **SUIT WEIGHT** Approximately 145 kg (320 lb)

- **LIFE SUPPORT DURATION** 8 hours (typical) with an extra 30 minutes back-up for emergencies

The chest-mounted display and control module features a manual temperature control for the spacesuit.

This battery-powered pistol grip tool loosens and tightens nuts and bolts.

The gloves are custom sewn for each astronaut for a perfect fit.

The coloured stripe identifies the lead astronaut on the spacewalk.

Suited and booted

The spacesuit includes helmets, gloves, boots, and an air supply. It takes an hour for the astronaut to put it on. The top half of the outer suit screws into the trousers to form an airtight seal. Oxygen is circulated throughout the suit for the astronaut to breathe. The helmet has headlights, while the gold-plated visor shields against the Sun's harmful rays.

FABRIC LAYERS

There are 14 layers in a spacesuit. The tough outer layer protects against space debris. This covers seven layers of aluminium-coated plastic film insulation, which keep temperatures constant inside the suit. The bladder layer holds oxygen to pressurize the suit, while the ripstop liners prevent tears.

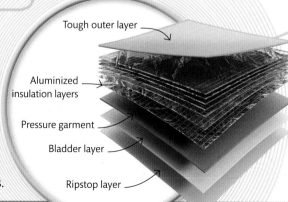

Tough outer layer

Aluminized insulation layers

Pressure garment

Bladder layer

Ripstop layer

The boots have soft soles to prevent damage to antennae and other delicate spacecraft parts.

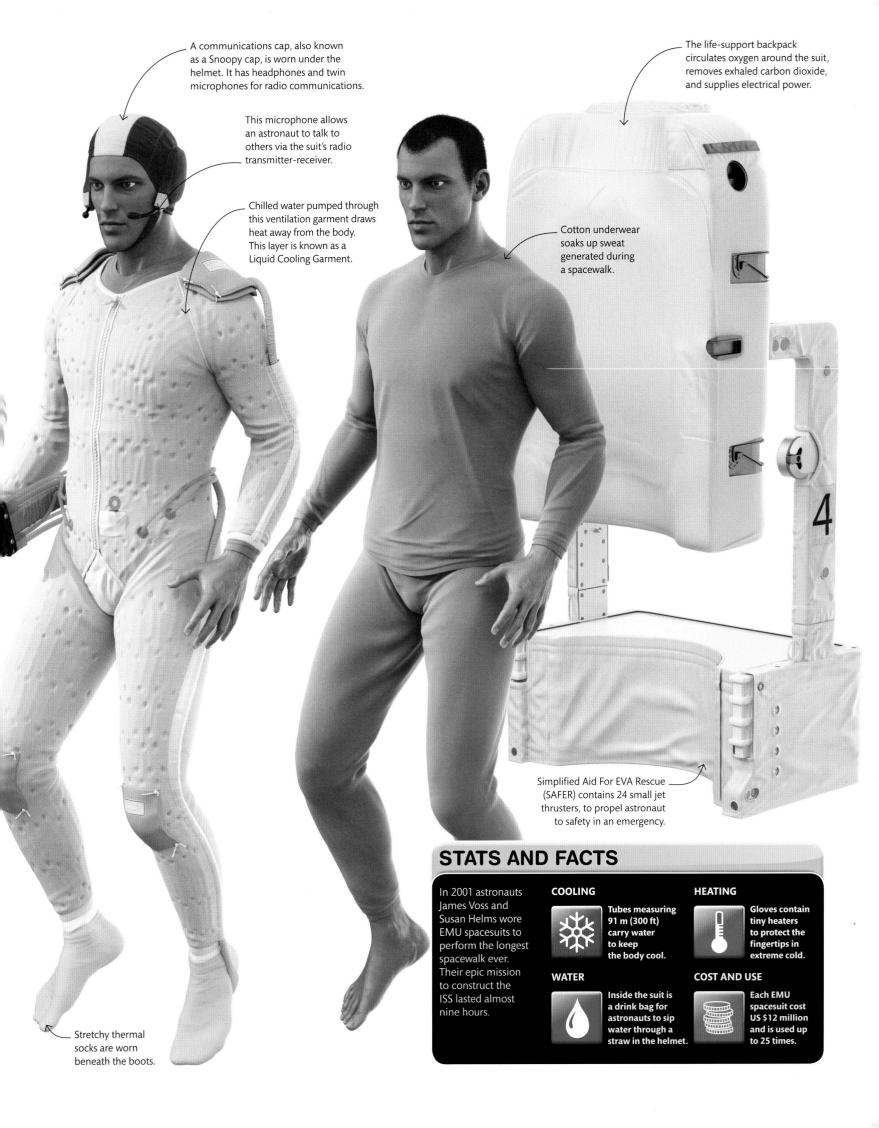

A communications cap, also known as a Snoopy cap, is worn under the helmet. It has headphones and twin microphones for radio communications.

This microphone allows an astronaut to talk to others via the suit's radio transmitter-receiver.

Chilled water pumped through this ventilation garment draws heat away from the body. This layer is known as a Liquid Cooling Garment.

The life-support backpack circulates oxygen around the suit, removes exhaled carbon dioxide, and supplies electrical power.

Cotton underwear soaks up sweat generated during a spacewalk.

Simplified Aid For EVA Rescue (SAFER) contains 24 small jet thrusters, to propel astronaut to safety in an emergency.

Stretchy thermal socks are worn beneath the boots.

STATS AND FACTS

In 2001 astronauts James Voss and Susan Helms wore EMU spacesuits to perform the longest spacewalk ever. Their epic mission to construct the ISS lasted almost nine hours.

COOLING
Tubes measuring 91 m (300 ft) carry water to keep the body cool.

HEATING
Gloves contain tiny heaters to protect the fingertips in extreme cold.

WATER
Inside the suit is a drink bag for astronauts to sip water through a straw in the helmet.

COST AND USE
Each EMU spacesuit cost US $12 million and is used up to 25 times.

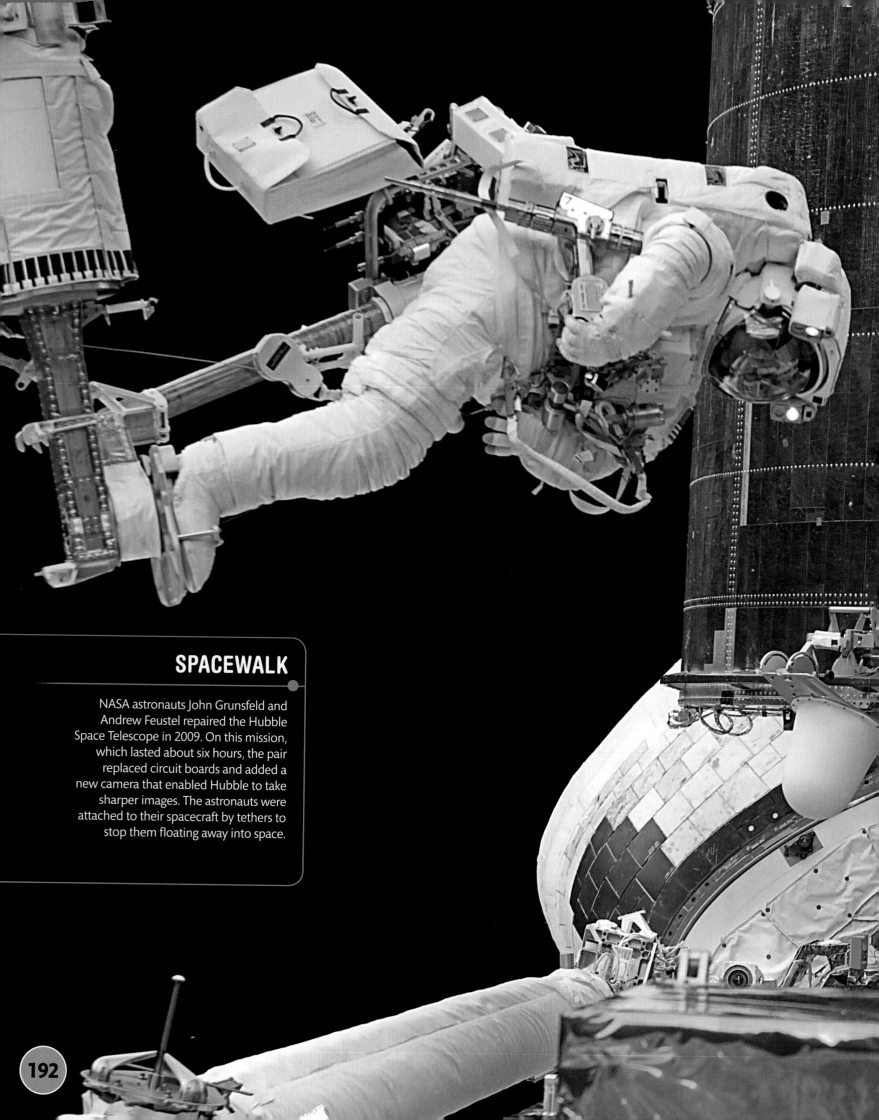

SPACEWALK

NASA astronauts John Grunsfeld and Andrew Feustel repaired the Hubble Space Telescope in 2009. On this mission, which lasted about six hours, the pair replaced circuit boards and added a new camera that enabled Hubble to take sharper images. The astronauts were attached to their spacecraft by tethers to stop them floating away into space.

CHINA'S MISSIONS
SHENZHOU

By the end of the 20th century, only the USA and Russia had built spacecraft capable of carrying people into space. The 21st century saw China become a space-faring nation with a new fleet named Shenzhou, which means "divine craft". In 2003, the Shenzhou 5 carried China's first astronaut, Yang Liewei, on 14 orbits around Earth. Six further Shenzhou missions followed, with five of them taking a small crew of astronauts in a craft similar to a super-sized Soyuz spacecraft. Each Shenzhou returned to Earth using a giant parachute measuring about three times the size of a basketball court.

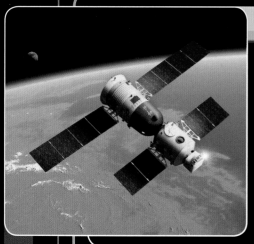

AT A GLANCE

- **LAUNCH VEHICLE** Long March 2F rocket

- **FIRST LAUNCH** 19 November 1999 (test flight without crew)

- **LAUNCH SITE** Jiuquan, northwestern China

- **ORGANIZATION** China National Space Administration (CNSA)

This artist's impression shows Shenzhou 5 orbiting Earth on China's first human spaceflight mission.

STATS AND FACTS

Shenzhou 9 carried China's first female astronaut, Liu Yang, into space on 16 June 2012, exactly 49 years to the day after Valentina Tereshkova became the first woman in space.

LENGTH

Shenzhou's three modules – service, re-entry, and orbital – have a total length of 9 m (30 ft).

m		5		10
ft	11	22		33

SOLAR PANELS

Four solar panels on Shenzhou generate a total of 1,500 watts of power.

SOFT LANDING

Six rockets fire when the re-entry module is 1 m (3 ft) above the ground to cushion its landing.

SPACECRAFT SIMULATOR

The crew of Shenzhou 9 is seen here in training onboard a simulator capsule that recreates the conditions when a spacecraft returns to Earth. In 2012, Shenzhou 9 became the fourth craft in China's fleet to be successfully launched. Controlled by computers on the ground, the craft carried this same three-person crew to dock at the Tiangong-1 space laboratory.

THE MOST POWERFUL
ROCKET
EVER BUILT

ROCKETING INTO THE FUTURE
SPACE LAUNCH SYSTEM

Currently under development, NASA's Space Launch System (SLS) will be the most powerful rocket ever built. It will be powered by two booster rockets similar to those once used to launch space shuttles, as well as two separate rocket stages, each with their own engines and fuel. The SLS will be able to launch either cargo or a capsule, called Orion, which will carry four astronauts. Eventually, it may launch missions to land on the Moon, visit asteroids, or even travel to Mars. The first launch, set for 2020, will send an Orion capsule without a crew to orbit around the Moon and then return.

AT A GLANCE

- **VEHICLE TYPE** Single-use, two-stage launch vehicle

- **FIRST LAUNCH** June 2020 (planned)

- **BOOSTER ROCKETS** The combined power of the two booster rockets is 44 million horsepower

- **ORGANIZATION** NASA

An engine to be used on the first SLS flight is being tested.

STATS AND FACTS

This heavy launch vehicle uses two separate stages and a total of four rocket engines. None of these are reusable because refurbishing is more costly than manufacturing new engines.

STANDING TALL

In its tallest configuration, the SLS will be 117 m (384 ft) tall.

BOOSTER BURN

Each SLS solid rocket booster burns 4,500 kg (10,000 lb) of fuel every second.

MAXIMUM PAYLOAD

The SLS will be able to carry cargoes of up to 130 tonnes, more than any other rocket.

MAXIMUM THRUST

The SLS's top thrust will be as much as 34 Jumbo Jets.

BLAST OFF

The SLS will take off from NASA's Kennedy Space Center on the coast of Florida, USA. Two external solid-rocket boosters will support the first stage engine in producing sufficient thrust at lift-off.

INFLATABLE SPACE HABITAT

B330

The B330 is an inflatable space habitat currently under development by Bigelow Aerospace in the USA. It expands to more than three times in size once inflated with air, whether in low Earth orbit or deep space, creating enough living space inside for six astronauts. Docking mechanisms allow the habitats to join other spacecraft or attach to each other in groups, forming space stations or even orbiting hotels.

Two large solar arrays generate electricity from sunlight.

The bridge area contains computers for navigation and communication with other spacecraft and ground control on Earth.

Propulsion engines move the habitat short distances through space for docking.

Space toilet and personal hygiene area

The galley equipment includes fixed ovens for heating food.

White bags lining the interior wall will hold fresh water or wastewater.

The crew rest area is behind soft fabric walls.

GALLEY

A galley kitchen area will contain at least one oven for cooking food, as well as hot water dispensers to heat pre-packed meals stored on board. Artificial lights will make it possible to grow plants for science experiments or to supply fresh food

AT A GLANCE

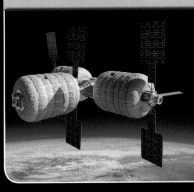

- **SPACECRAFT TYPE** Inflatable habitat
- **SIZE** 16.9 m × 6.7 m (55 ft × 22 ft)
- **LAUNCH DATE** 2021 (proposed)
- **ORGANIZATION** Bigelow Aerospace

Here three B330s have joined to form a space station.

BEAM

In 2016, a small inflatable module called BEAM (the Bigelow Expandable Activity Module) docked to the International Space Station (ISS) in a trial devised to test the design and materials used in B330. BEAM proved to be just as resilient as traditional metal modules.

EVA suits are available for astronauts to conduct spacewalks outside the habitat.

Storage bags and lockers are fitted to the interior wall.

Soft fabric walls can be repositioned around the habitat to section off parts of the living and working space.

Long-term storage containers hold supplies and spare parts.

Versatile habitat

This artist's impression of a B330 shows the large interior divided into separate living, working, and storage areas. Grab rails around the habitat allow astronauts to move around in the weightlessness of space. Flexible partitioning allows the habitat to be adapted for different purposes, such as scientific research or space tourism.

STATS AND FACTS

When B330 is fully inflated, its outer shell is as hard as concrete. The wall is 46 cm (18 in) thick and consists of 20 different layers, which together protect the interior from extreme temperature swings and harmful radiation.

WEIGHT

The B330 weighs about 23 tonnes, which is 20 times lighter than the ISS.

DOCKING

Onboard docking systems enable B330 to connect to different types of spacecraft.

ORBITING THE MOON

A B330 habitat is planned to be placed in lunar orbit in 2022.

VIEW

Windows coated with ultraviolet protection film give astronauts a view of space.

THE FIRST INFLATABLE SPACE HABITAT

GLOSSARY

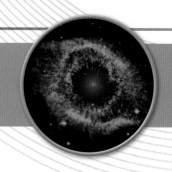

ACCRETION DISC
A swirling disc of material around a star or black hole.

ACTIVE GALAXY
A galaxy that emits an exceptional amount of energy, much of which comes from a supermassive black hole in its centre.

ANTENNA
A rod- or dish-like structure on spacecraft and telescopes used to transmit and receive radio signals.

ASTEROID
A small, irregular Solar System object, made of rock and/or metal, that orbits the Sun.

ASTEROID BELT
A doughnut-shaped region of the Solar System, between the orbits of Mars and Jupiter, that contains a large number of orbiting asteroids.

ASTRONAUT
A person trained to travel and live in space.

ASTRONOMER
A person who studies the stars and planets and other objects in space.

ATMOSPHERE
The outer layer of gas around a planet or star.

ATOM
The smallest particle of a chemical element that can exist on its own.

AURORA
A display of glowing gas in a planet's upper atmosphere, most often over its polar regions.

AXIS
The imaginary line that passes through the centre of a planet or star and around which the planet or star rotates.

BARRED SPIRAL GALAXY
A galaxy with spiral arms that curl out from the ends of a bar-shaped nucleus.

BIG BANG
The event in which the Universe was created 13.8 billion years ago. According to the Big Bang theory, the Universe began in an extremely dense and hot state and has been expanding ever since. The Big Bang was the origin of space, time, and matter.

BINARY STAR
A pair of stars orbiting around a common centre due to their mutual gravity.

BLACK DWARF
A cold dead star that forms when a white dwarf slowly cools down over billions of years.

BLACK HOLE
An object in space with such a strong gravitational pull that nothing, not even light, can escape from it.

BLAZAR
An active galaxy with a supermassive black hole at its centre.

BROWN DWARF
An object in space with a size range between that of a massive planet and small stars. They are sometimes known as "failed stars".

CHARGED PARTICLE
A particle that has a positive or negative electrical charge.

CHROMOSPHERE
The layer of the Sun's atmosphere above the photosphere.

CIRCUIT
An unbroken path that allows electricity to flow along it.

CLIFF
A steep rock face that rises high from the surface and is almost vertical, like a wall.

COMET
An object made of dust and ice that travels around the Sun in an elliptical orbit. As it gets near the Sun, the ice starts to vaporize, creating tails of dust and gas.

COMETARY KNOTS
When fast-moving, recently released gas interacts with denser gas surrounding a star, comet-like knots form.

CONSTELLATION
A named area of the sky as defined by the International Astronomical Union (IAU). The sky is divided into 88 constellations.

CONVECTIVE ZONE
A layer inside a star such as the Sun where hot gas rises, cools, and then sinks down again. This movement carries energy from the core to the surface.

CORONA
The upper atmosphere of the Sun. It is made of the same gases as the rest of the star, but it is less dense and much hotter than the Sun's surface. The corona extends far beyond the Sun, reaching millions of kilometres into space.

COSMONAUT
A Russian astronaut.

COSMOS
The Universe is also referred to as the cosmos.

CRATER
A bowl-shaped hollow on the surface of a planet or moon.

CRUST
The thin, solid outer layer of a planet or moon.

DENSITY
The amount of matter an object has in relation to its volume.

DEPLOY
To send into action.

DOCKING
When one spacecraft meets another in space and they connect together.

DWARF GALAXY
A small galaxy typically containing only a million to several billion stars.

DWARF PLANET
A small Solar System body that has a spherical shape like a planet but orbits the Sun in the Asteroid Belt or the Kuiper Belt. Currently, five objects are designated as dwarf planets: Ceres in the Asteroid Belt, and Pluto, Eris, Haumea, and Makemake in the Kuiper Belt.

ECLIPSE
An astronomical event in which an object either passes into the shadow of another object or temporarily blocks an observer's view. During a solar eclipse, the shadow of the Moon falls on Earth. In a lunar eclipse, the shadow of Earth falls on the Moon.

ELECTROMAGNETIC RADIATION
Energy waves that can travel through space and matter. Visible light, X-rays, and microwaves are all forms of electromagnetic radiation.

ELLIPTICAL GALAXY
In this oval-shaped galaxy the stars are very old and the galaxy doesn't contain much gas or dust.

EQUATOR
The imaginary line around a planet, halfway between its north and south poles.

EXOPLANET
A planet that orbits a star other than the Sun.

EXTRAVEHICULAR ACTIVITY (EVA)
An activity performed by an astronaut outside a spacecraft in space or on the surface of a moon or planet.

EXTRAVEHICULAR MOBILITY UNIT (EMU)
An International Space Station suit that includes a Primary Life Support System (PLSS) backpack, which controls both the temperature and pressure, and supplies oxygen.

FLYBY
The flight of a spacecraft, which passes a planet without attempting to land or to orbit it.

GALAXY
A collection of millions or trillions of stars, gas, and dust held together by gravity.

GAMMA RAYS
Electromagnetic radiation that has a very short wavelength and is the most powerful of all.

GAS GIANT
A planet that is made mostly of hydrogen and helium, such as Jupiter or Saturn.

GEYSER
Hot spring that spurts a column of water and steam into the air.

GLOBULAR CLUSTER
A ball-shaped cluster of stars that orbit a large galaxy.

GRAVITY
The force that pulls all objects that have mass and energy towards one another. It is the force that keeps moons in orbit around planets, and planets in orbit around the Sun.

HABITABLE ZONE
The region around a star where life could develop.

HABITAT
The natural environment where plants, animals, and other organisms live.

HEAT SHIELD
A device fitted to a spacecraft that protects astronauts or equipment from extreme heat, such as that experienced when re-entering the Earth's atmosphere.

HEMISPHERE
One half of a sphere. Earth is divided into northern and southern hemispheres by the equator.

HYDROTHERMAL
Relating to water that has been naturally heated under ground.

HYPERGIANT
The most massive stars in the Universe today are known as hypergiants. They are thought to have the mass of more than 100 Suns and the luminosity of more than a million Suns.

ICE GIANT
A planet like Uranus or Neptune with a small rocky core, covered by a thick liquid layer mainly of water, ammonia, and methane, and an atmosphere of hydrogen and helium.

IMPACT CRATER
A basin-like depression caused by the crash of an object falling from space; the displaced material usually forms a raised rim around it.

INFRARED
Electromagnetic radiation with wavelengths shorter than radio waves but longer than visible light. Infrared radiation can be felt as heat.

INTERACTING GALAXIES
Two galaxies that come so close to each other that their gravity disrupts the dust and gas around them, triggering star formation.

IRREGULAR GALAXY
A galaxy with no obvious shape or structure. They contain lots of young stars, gas, and dust.

JETTISON
The act of releasing or throwing away something, such as used parts of a multistage rocket.

KUIPER BELT
A region in the Solar System beyond Neptune where a large number of comets and icy asteroids are located.

LASER BEAM/LIGHT
A narrow beam of intense light at a single wavelength.

LAUNCH VEHICLE
A rocket-powered vehicle that is used to send spacecraft or satellites into space.

LENS
A transparent glass disc with at least one curved surface. A telescope lens is usually made up of a number of separate lenses or groups of lens elements, which can be adjusted to bring an image into focus.

LENTICULAR GALAXY
A galaxy in the shape of a convex lens. It has a central bulge but no spiral arms.

LIGHT YEAR
The distance travelled by light in a vacuum in one year.

LOCAL GROUP
A small group of just over two dozen galaxies of which the Milky Way galaxy is a member.

LUMINOSITY
The total amount of energy emitted in one second by a star or galaxy.

MAGNETIC FIELD
The region surrounding a magnetized object or moving charged particles, where other magnetized objects or charged particles feel a force of attraction or repulsion.

MAGNETOMETER
An instrument used for measuring magnetic forces.

MAGNITUDE
The brightness of a celestial object, which can be measured in two ways. An object's apparent magnitude is how bright it appears in the night sky when viewed from Earth. Its absolute magnitude, or luminosity, is the amount of light energy emitted by the object.

MAIN SEQUENCE STAR
An ordinary star, such as our Sun, that shines by converting hydrogen to helium.

MARIA
Large, flat areas on the Moon's surface that look dark when viewed from Earth. These areas were originally thought to be lakes or seas but are now known to be plains of solidified lava.

METEOR
A streak of light, also called a shooting star, seen when a meteoroid burns up due to friction on entering Earth's atmosphere.

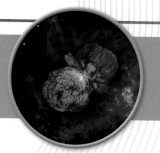

METEORITE
A meteoroid that reaches the ground and survives impact. Meteorites are usually classified according to their composition as stony, iron, or stony-iron.

METEOROID
A particle of rock, metal, or ice travelling through space.

METEOR SHOWER
Meteors with trails that can all be traced back to the same point in the sky and are seen over a period of a few hours or days.

MICROGRAVITY
The condition of weightlessness experienced by astronauts when in orbit or in free-fall.

MILKY WAY
The barred spiral galaxy that contains the Solar System and is visible to the naked eye as a band of faint light across the night sky.

MINERAL
Naturally occurring solid made of crystals. Minerals are made up of specific combinations of elements.

MODULE
A portion of a spacecraft.

NEBULA
A cloud of gas and/or dust in space.

NEUTRON STAR
A dense collapsed star that is mainly made of neutrons.

NUCLEAR FUSION
A process in which two atomic nuclei join to form a heavier nucleus and release large amounts of energy.

NUCLEUS
The compact central core of an atom. Also the solid, icy body of a comet.

OBSERVATORY
A place where scientists observe astronomical objects and radiation from space with telescopes or other instruments. The two main kinds are optical and radio observatories. Robotic observatories orbiting in space, such as X-ray observatories, are controlled remotely from the ground.

OORT CLOUD
A huge spherical cloud, about 1.6 light years wide, that surrounds the Sun and planets. It contains billions of comets.

OPEN STAR CLUSTER
A loose star cluster formed when a group of stars is born inside a nebula.

ORBIT
The path taken by an object around another when affected by its gravity. The orbits of planets are elliptical in shape.

ORBITER
A spacecraft that is designed to orbit an object but not land on it.

PAYLOAD
Cargo or equipment carried into space by a rocket or a spacecraft.

PERISCOPE
A device that uses mirrors to allow the user to see things that are not in his or her direct line of sight.

PHOTOSPHERE
The thin gaseous layer at the base of the Sun's atmosphere from which visible light is emitted.

PLANET
A spherical object that orbits a star and is sufficiently massive to have cleared its orbital path of debris.

PLANETARY NEBULA
A glowing cloud of gas around a star at the end of its life.

PLANETESIMALS
Small rocky or icy objects formed in the early Solar System that were pulled together by gravity to form planets.

PLATES
Earth's crust is made up of a number of gigantic slabs called plates, which float on the surface of the layer beneath.

PLUME
A tall column of gas, liquid, or small particles that rises from beneath the surface of a moon or planet through a crack or vent.

PROBE
A spacecraft that investigates a planet or moon's atmosphere or surface and is usually transported by a larger spacecraft.

PROMINENCE
A large, flame-like plume of plasma (a form of hot gas consisting of electrically charged particles) emerging from the Sun's photosphere.

PROTOPLANET
A planet in its early stage of formation.

PROTOSTAR
A star in its early stage of formation.

PULSAR
A neutron star that sends out beams of radiation as it spins.

QUASAR
Short for "quasi-stellar radio source", a quasar is the immensely luminous nucleus of a distant active galaxy with a supermassive black hole at its centre.

RADAR
A detection system that bounces radio waves off an object and collects the reflected signals to locate an object and/or map its shape.

RADIANT
The place in the sky towards which the trails of meteors belonging to the same shower can be traced back.

RADIATION
The transfer of energy as waves or particles through space.

RADIO ASTRONOMER
A person who studies celestial objects by analysing radio waves from space.

RADIO WAVES
A type of electromagnetic radiation with wavelengths longer than infrared light.

RED GIANT
A large, luminous star with a low surface temperature and a reddish colour. It burns helium in its core rather than hydrogen and is nearing the final stages of its life.

RE-ENTRY
The descent of a spacecraft into Earth's atmosphere from space.

RESOLUTION
The ability of a lens or imaging device to record fine detail.

RETROROCKET
A rocket system used for slowing a spacecraft down rather than accelerating it. Retrorockets are used to begin re-entry to Earth's atmosphere, or to slow space probes down when they arrive at their destination.

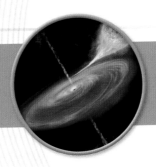

ROVER
A vehicle that is driven remotely on the surface of a planet and moon.

SATELLITE
A natural body orbiting another more massive body, or an artificial body orbiting Earth.

SCARP
A long, steep slope or cliff between two areas of land at different heights.

SEYFERT GALAXY
A spiral-shaped active galaxy with an exceptionally luminous and compact nucleus.

SOLAR NEBULA
A vast, spinning cloud of gas and dust from which everything in the Solar System – the Sun, planets, moons, and smaller objects – formed.

SOLAR WIND
A continuous flow of fast-moving charged particles from the Sun.

SPACE
The region beyond Earth's atmosphere, in which all bodies of the Universe exist. Also used to describe the voids between space bodies, such as the planets, stars, and galaxies.

SPACECRAFT
A vehicle that travels in space.

SPACE RACE
The 20th-century space-related competition between the Soviet Union and the United States. It started with the launch of Sputnik 1 and resulted in putting the first humans on the Moon.

SPACESUIT
The all-in-one sealed clothing unit worn by astronauts when outside their craft in space.

SPECTROGRAPH
A device to split light, or other radiation, into a spectrum that can be measured and analysed.

SPEED OF LIGHT
The distance travelled by light through empty space in one second – nearly 300,000 km. Nothing can travel faster than this speed.

SPIRAL GALAXY
A disc-shaped galaxy with spiral arms.

STARBURST GALAXY
A galaxy that has undergone a sudden period of star formation, often as the result of colliding with another galaxy.

STAR CLUSTER
A group of stars held together by gravity. Open clusters are loose groups of a few hundred young stars; globular clusters are dense balls containing many thousands of old stars.

STELLAR
Relating to stars.

SUB-SURFACE OCEAN
An ocean that lies below the surface of rock or ice.

SUNSPOT
A region of intense magnetic activity in the Sun's photosphere that appears darker than its surroundings.

SUPERGIANT
An exceptionally large and luminous star. Stars at least 10 times as massive as the Sun swell into supergiants at the end of their lives.

SUPERNOVA
A dying star that explodes and gives off so much energy it can shine brighter than a whole galaxy.

THRUST
The force from an engine that propels a rocket or spacecraft forwards.

THRUSTER
A small rocket engine on a spacecraft, or a secondary jet or propeller on a ship or underwater craft.

ULTRAVIOLET RADIATION (UV)
Electromagnetic radiation with wavelengths shorter than visible light but longer than X-rays.

UNIVERSE
Everything that exists: all the galaxies, stars, and planets, and the space in between, including all things on Earth.

VARIABLE STAR
A star that changes in brightness. Many variable stars also regularly change size.

VISIBLE LIGHT
The part of the electromagnetic spectrum that we can see with our eyes.

WATER VAPOUR
The gaseous form of water.

WAVELENGTH(S)
The distance between two peaks or troughs of a wave.

WEATHERING
The breakdown of rock by the weather.

WEATHER SATELLITE
A device that orbits Earth and sends back data to help scientists forecast the weather.

WEIGHTLESSNESS
The sensation of constantly floating that astronauts experience when they are travelling in space.

WHITE DWARF
A small, dim star that has stopped generating energy by nuclear reaction. When a star reaches its white dwarf stage, it is close to the end of its life.

X-RAYS
Electromagnetic radiation with wavelengths shorter than ultraviolet radiation but longer than gamma rays.

Abbreviations used in this book	
/	per – for example, km/h means kilometres per hour
cm	centimetre
cu m	cubic metre
cu ft	cubic feet
°C	degrees Celsius
°F	degrees Fahrenheit
ft	foot
GB	gigabyte
h	hour
in	inch
kg	kilogram
km	kilometre
lb	pound
m	metre
mph	miles per hour
sq	square

INDEX

ACKNOWLEDGMENTS

Dorling Kindersley would like to thank the following people:
Sam Atkinson, Bharti Bedi, Ankona Das, Arpita Dasgupta, Tina Jindal, Andrea Mills, and Pauline Savage for editorial assistance; Srishti Arora for design assistance; Deepak Mittal for DTP assistance; Ed Aves for proofreading; and Carron Brown for indexing.

Picture Credits
The publisher would like to thank the following for their kind permission to reproduce their photographs:

(Key: a-above; b-below/bottom; c-centre; f-far; l-left; r-right; t-top)

1 Alamy Stock Photo: Andrey Armyagov. **2-3 Ian Regan**. **4 ESA:** ESA / DLR / FU Berlin (G. Neukum),CC BY-SA 3.0 IGO (ca). **NASA and The Hubble Heritage Team (AURA/STScI):** NASA, ESA, and the Hubble Heritage Team (STScI / AURA) (cla). **NASA:** NASA / JPL-Caltech / SwRI / MSSS / Gerald Eichstadt / Sean Doran (ca/redspot); NASA / CXC / SAO (cra). **5 Alamy Stock Photo:** Science History Images (cra). **NASA:** NASA / Johns Hopkins APL / Steve Gribben (ca). **Science Photo Library:** NASA / CXC / UNIVERSITY OF HERTFORDSHIRE / M. HARDCASTLE ET AL / CSIRO / ATNF / ATCA (cla). **10 Science Photo Library:** Luis Argerich (cla). **12 ESA / Hubble:** NASA, ESA and the Hubble Heritage (STScI / AURA)-ESA / Hubble Collaboration. Acknowledgment: M. West (ESO, Chile) (clb). **NASA and The Hubble Heritage Team (AURA/STScI):** NASA, ESA, and The Hubble Heritage Team (STScI / AURA); P. Knezek (WIYN) (cla). **12-13 Robert Gendler**. **13 ESA / Hubble:** ESA / Hubble & NASA (tl). **NASA:** ASA, ESA, A. Riess (STScI / JHU), L. Macri (Texas A&M University), and Hubble Heritage Team (STScI / AURA) (cr); NASA / JPL-Caltech / STScI (tr); NASA / JPL-Caltech (crb). **16 NASA:** NASA / JPL-Caltech / UCLA (cla). **17 Alamy Stock Photo:** Science History Images (cra). **ESA:** JAXA (cla). **NASA:** NASA / CXC / SAO / DSS / D. Patnaude (clb); X-ray: NASA / CXC / SAO; Optical: NASA / STScI (c); NASA / JPL-Caltech / STScI / CXC / SAO (crb). **NRAO:** NRAO / AUI / NSF; D. Berry / Skyworks (bc). Science Photo Library: Russell Kightley (cb). **18-19 NASA and The Hubble Heritage Team (AURA/ STScI):** NASA, ESA, and the Hubble Heritage Team (STScI / AURA). **20 Science Photo Library:** Mark Garlick (cl). **20-21 ESA:** ESO / L. Calçada. **21 Science Photo Library:** Mark Garlick (tr); Mark Garlick (crb). **24 Alamy Stock Photo:** Zachary Frank (c). **NASA and The Hubble Heritage Team (AURA/STScI):** JPL / Caltech (cb); NRAO / AUI / NSF (cl). **iStockphoto.com:** Stocktrek Images (bc). **24-25 NASA and The Hubble Heritage Team (AURA/STScI):** NASA, ESA, G. Dubner (IAFE, CONICET-University of Buenos Aires) et al; A. Loll et al; T. Temim et al; F. Seward et al; VLA / NRAO / AUI / NSF; Chandra / CXC; Spitzer / JPL-Caltech; XMM-Newton / ESA; and Hubble / STScI. **25 ESA / Hubble:** NASA / Chandra / CXC (c); ESA (tc). **ESA:** ESA-C. Carreau (cra). **NASA. Science Photo Library:** NASA (br). **26 Alamy Stock Photo:** Everett Collection Inc (cl). **ESA:** ESA / ATG medialab (bc). **NASA. Rex by Shutterstock:** JAXA (clb). **Science Photo Library:** Detlev Van Ravenswaay (ca); Detlev Van Ravenswaay (c); SPUTNIK (cra). **27 Alamy Stock Photo:** ITAR-TASS News Agency (cra); SPUTNIK (cl). **Getty Images:** Space Frontiers (bc); Keystone-France (cla). **NASA:** NASA / Neil Armstrong (c). **National Air and Space Museum, Smithsonian Institution:** (cb). **Science Photo Library:** Detlev Van Ravenswaay (clb); SPUTNIK (br). **28-29 NASA**. **32 Dorling Kindersley:** Peter Bull / NASA / ESA (cl). **Getty Images:** SinghaphanAll (bl). **NASA:** NASA / SDO (cb). **33 Science Photo Library:** Scharmer et al, Royal Swedish Academy Of Sciences (tr). **34-35 NASA:** NASA / SDO. **36-37 NASA:** NASA / Johns Hopkins University Applied Physics Laboratory / Carnegie Institution of Washington. **36 NASA:** NASA / Johns Hopkins University Applied Physics Laboratory / Carnegie Institution of Washington (cl). **38-39 NASA:** NASA / JPL. **38 NASA:** NASA / JPL (cl). **40 NASA:** NASA Goddard Space Flight Center Image by Reto Stöckli (land surface, shallow water, clouds). Enhancements by Robert Simmon (ocean color, compositing, 3D globes, animation). Data and technical support: MODIS Land Group; MODIS Science Data Support Team; MODIS Atmosphere Group; MODIS Ocean Group Additional data: USGS EROS Data Center (topography); USGS Terrestrial Remote Sensing Flagstaff Field Center (Antarctica); Defense Meteorological Satellite Program (city lights) (cl). **41 Alamy Stock Photo:** Buiten-Beeld (br); Sergi Reboredo (cr). Getty Images: Mint Images - Art Wolfe (cra); Vincent Pommeyrol (crb). **42-43 NASA:** NASA Earth Observatory image by Robert Simmon, using Suomi NPP VIIRS data provided courtesy of Chris Elvidge (NOAA National Geophysical Data Center). Suomi NPP is the result of a partnership between NASA, NOAA, and the Department of Defense. **44 NASA:** NASA / GSFC / Arizona State University (cl); NASA / Goddard / Arizona State University. (cl/farsaide). **45 Lunar and Planetary Institute:** NASA (br). **46-47 NASA:** NASA / Goddard / Arizona State University. **48 ESA:** ESA / DLR / FU Berlin (G. Neukum),CC BY-SA 3.0 IGO (cl). **49 NASA:** NASA / JPL-Caltech / University of Arizona (cra); NASA / JPL-Caltech / University of Arizona (crb); NASA / JPL-Caltech / University of Arizona (br). **50-51 NASA:** NASA / JPL / USGS. **52-53 Getty Images:** Carlos Fernandez. **52 Getty Images:** Jeff Dai / Stocktrek Images (cl). **54-55 Dreamstime**.com: Galyna Andrushko. **54 Dreamstime.com:** Pytyczech (cl). **56-57 NASA:** NASA / JPL-Caltech / UCLA / MPS / DLR / IDA / PSI. **56 NASA:** NASA / JPL-Caltech / UCLA / MPS / DLR / IDA (cl). **60 NASA:** NASA / JPL-Caltech / SwRI / ASI / INAF / JIRAM (cl). **61 NASA:** NASA / JPL-Caltech / SwRI / MSSS / Gerald Eichstadt / Sean Doran (tc); NASA / ESA, The Hubble Heritage Team. Acknowledgment: H. Weaver (JHU / APL) and A. Simon-Miller (NASA / GSFC (bc). **62-63 NASA:** NASA / JPL-Caltech / SwRI / MSSS / Gerald Eichstadt / Sean Doran. **64 NASA:** NASA / JPL / University of Arizona (cl). **65 Dreamstime.com:** Vjanez (t/all). **66-67 Ian Regan**. **68-69 Science Photo Library:** Detlev Van Ravenswaay. **71 NASA:** NASA / JPL (cr). **72-73 Adobe Stock:** Supermurmel. **73 NASA:** NASA / JPL (cr). **74-75 Dr. Dominic Fortes, UCL**. **76-77 ESO:** S. Deiries / ESO. **77 UCL Observatory/Steve Fossey/Stephen Boyle (https://creativecommons.org/licenses/by/3.0/):** (cr). **78-79 NASA:** NASA / JHUAPL / SwRI. **78 NASA:** NASA / JHUAPL / SwRI (cl). **82-83 ESA / Hubble:** Akira Fujii. **83 ESA / Hubble:** NASA, ESA, H. Bond (STScI), and M. Barstow (University of Leicester) (cr). **86 NASA:** NASA, ESA, and F. Paresce (INAF-IASF, Bologna, Italy), R. O'Connell (University of Virginia, Charlottesville), and the Wide Field Camera 3 Science Oversight Committee (cl). **88-89 ESA / Hubble:** ESO / IDA / Danish 1.5 m / R.Gendler, J-E. Ovaldsen, C. Thöne, and C. Feron. **89 Alamy Stock Photo:** S.E.A. Photo (crb). **90 ESA / Hubble:** A. Fujii (cl). **90-91 ESO:** ESO / INAF-VST / OmegaCAM. Acknowledgement: A. Grado, L. Limatola / INAF-Capodimonte Observatory. **92-93 Marco Lorenzi (https://www.glitteringlights.com)**. **92 Till Credner / AlltheSky.com**. **94 ESA / Hubble:** NASA / ESA / Hubble / F. Ferraro (cl). **94-95 Science Photo Library:** Russell Kightley. **95 Alamy Stock Photo:** Jim West (cra). **96 Till Credner / AlltheSky.com:** (clb). **96-97 NASA:** NASA / CXC / SAO. **98 NASA**. **100-101 Getty Images:** Stocktrek Images. **101 ESA / Hubble:** NASA, C.R. O'Dell and S.K. Wong (Rice University) (crb). **102 ESA / Hubble:** NASA, ESA, N. Smith (University of California, Berkeley), and The Hubble Heritage Team (STScI / AURA) (cl). **102-103 ESA / Hubble:** NASA, ESA, and M. Livio, The Hubble Heritage Team and the Hubble 20th Anniversary Team (STScI). **104-105 NASA:** NASA / JPL-Caltech / University of Arizona. **105 NASA and The Hubble Heritage Team (AURA/STScI):** NASA, NOAO, ESA, the Hubble Helix Nebula Team, M. Meixner (STScI), and T.A. Rector (NRAO). (crb). **106 NASA:** NASA / Ames / SETI Institute / JPL-Caltech (cla); NASA / Ames / SETI Institute / JPL-Caltech (c). **107 NASA:** NASA / Ames / SETI Institute / JPL-Caltech (c). **108-109 ESO:** ESO / M. Kornmesser. **112 NASA:** NASA, ESA, and Hubble Heritage Team (STScI / AURA, Acknowledgment: T. Do, A.Ghez (UCLA), V. Bajaj (STScI). **112-113 NASA:** NASA / JPL-Caltech / R.Hurt (SSC-Caltech). **113 NASA**. **114-115 Mark Gee / theartofnight.com**. **116-117 NRAO:** NRAO / AUI / NSF; D. Berry / Skyworks. **116 NASA:** NASA / CXC / MIT / F.K. Baganoff et al. (clb). **118-119 Robert Gendler**. **118 NASA and The Hubble Heritage Team (AURA/STScI):** NASA, ESA, Z. Levay and R. van der Marel (STScI), T. Hallas, and A. Mellinger (cl). **120-121 NASA and The Hubble Heritage Team (AURA/STScI):** NASA, ESA, and the Hubble Heritage Team (STScI / AURA). **120 Alamy Stock Photo:** Science History Images (cl). **122-123 ESA / Hubble:** NASA, ESA, S. Beckwith (STScI), and The Hubble Heritage Team (STScI / AURA). **123 ESA / Hubble:** H. Ford (JHU / STScI), the Faint Object Spectrograph IDT, and NASA / ESA (crb). **124 ESA / Hubble:** NOAO / AURA / NSF, B. Twardy, B. Twardy, and A. Block (NOAO) (clb). **124-125 ESA / Hubble:** ESA / Hubble & NASA. **126-127 ESO:** ESO / M. Kornmesser. **126 ESO:** ESO / UKIDSS / SDSS (clb). **127 ESA / Hubble:** ESA / Hubble & NASA (cr). ESO: ESO / L. Calçada (crb). **NASA:** NRAO (br). **Science Photo Library:** NASA / CXC / UNIVERSITY OF HERTFORDSHIRE / M. HARDCASTLE ET AL / CSIRO / ATNF / ATCA (cra). **128 Dan Crowson:** (clb). **128-129 ESA / Hubble: NASA / ESA and The Hubble Heritage Team (STScI / AURA)**. **132-133 Getty Images:** Joe McNally. **133 Getty Images:** Roger Ressmeyer / Corbis / VCG (crb). **134-135 ESO:** ESO / L. Calçada. **134 Dorling Kindersley:** ESO (clb). **136-137 Getty Images:** VCG. **137 Alamy Stock Photo:** Xinhua (cr). **138-139 Getty Images:** © Roger Ressmeyer / Corbis / VCG. **139 Getty Images:** Raphael GAILLARDE (crb). **140 NASA**. **141 Getty Images:** Historical (tc). **NASA** (cra, cr). **142-143 ESA / Hubble:** NASA, ESA, STScI. **144 NASA:** NASA / CXC / Rutgers / G.Cassam-Chenai, J.Hughes et al; Radio: NRAO / AUI / NSF / GBT / VLA / Dyer, Maddalena & Cornwell; Optical: Middlebury College / F. Winkler, NOAO / AURA / NSF / CTIO Schmidt & DSS (clb). **144-145 NASA:** NASA / CXC / NGST. **146 NASA:** NASA / Desiree Stover (clb). **146-147 NASA:** NASA, ESA, and Northrop Grumman. **148-149 NASA:** NASA / JHUAPL / Carnegie Institution of Washington. **148 NASA:** NASA / Johns Hopkins University Applied Physics Laboratory / Carnegie Institution of Washington (clb). **149 NASA**. **150-151 NASA:** NASA / Johns Hopkins APL / Steve Gribben. **150 NASA**. **152-153 NASA:** NASA / JPL-Caltech / MSSS. **153 NASA:** NASA / JPL-Caltech / MSSS (crb). **154 NASA:** NASA / JPL-Caltech / UCLA / MPS / DLR / IDA (cb). **154-155 NASA:** NASA / JPL-Caltech. **156-157 ESA:** ESA / ATG medialab. **157 ESA:** ESA / Rosetta / NAVCAM - CC BY-SA IGO 3.0 (crb). **158-159 NASA:** NASA / JPL-Caltech. **158 NASA:** NASA / JPL-Caltech / SwRI / JunoCam (clb). **160-161 NASA:** NASA / JPL-Caltech. **161 NASA:** ESA-D. Ducros (crb). **162-163 Science Photo Library:** Carlos Clarivan. **163 NASA:** NASA / JPL-Caltech (cr). **164-165 NASA:** NASA / JHUAPL / SwRI. **164 NASA**. **165 NASA**. **166-167 Japan Aerospace Exploration Agency (JAXA):** (All). **171 Alamy Stock Photo:** Eduardo Fuster Salamero (tl); SPUTNIK (cr); SPUTNIK (br). Science Photo Library: SPUTNIK (bc). **172 NASA:** NASA / Aubrey Gemignani (cl). **172-173 Alamy Stock Photo:** Andrey Armyagov. **173 ESA:** GCTC (tc). NASA: (cb). **174-175 NASA**. **176 Alamy Stock Photo:** NASA Archive (clb). **176-177 Dreamstime.com:** Wangkun Jia. **178 Dorling Kindersley:** NASA (cb). **179 Dorling Kindersley:** Andy Crawford / Bob Gathany (tr). **180-181 Alamy Stock Photo:** Science History Images. **182-183 NASA**. **183 NASA:** Lori Losey / NASA (crb). **184-185 NASA:** NASA / Ken Ulbrich. **186 NASA**. **187 NASA**. **188-189 NASA**. **190 NASA**. **192-193 NASA**. **194 Science Photo Library:** Detlev Van Ravenswaay (clb). **194-195 Rex by Shutterstock:** Qin Xian'An / Chine Nouvelle / Sipa. **196-197 Google Earth**. **197 NASA**. **199 NASA:** Bigelow Aerospace (tr). **200 ESO:** ESO / L. Calçada (tr). **NASA:** NASA / JPL-Caltech / University of Arizona (tr/helix). **201 ESA / Hubble:** NASA / Chandra / CXC (tr). **202 Alamy Stock Photo:** S.E.A. Photo (tc). **ESA / Hubble:** NASA, ESA, and M. Livio, The Hubble Heritage Team and the Hubble 20th Anniversary Team (STScI) (tc/carina); NASA, ESA, S. Beckwith (STScI), and The Hubble Heritage Team (STScI / AURA) (tr). **203 Alamy Stock Photo:** Science History Images (tr). **NASA:** (tl); NASA / JPL-Caltech / UCLA / MPS / DLR / IDA / PSI (tc). **204 Getty Images:** Stocktrek Images (tc). **Marco Lorenzi (https://www.glitteringlights.com):** (tr). **NASA:** NASA / JPL (tc/venus). **ESA:** ESO / L. Calçada (tl). **NASA:** Lori Losey / NASA (tc/747); NASA / JPL-Caltech (tc). **208 NASA**

Endpaper images: Front: **ESA / Hubble:** NASA, ESA, N. Smith (University of California, Berkeley), and The Hubble Heritage Team (STScI / AURA); Back: **ESA / Hubble:** NASA, ESA, N. Smith (University of California, Berkeley), and The Hubble Heritage Team (STScI / AURA)

For further information see:
www.dkimages.com

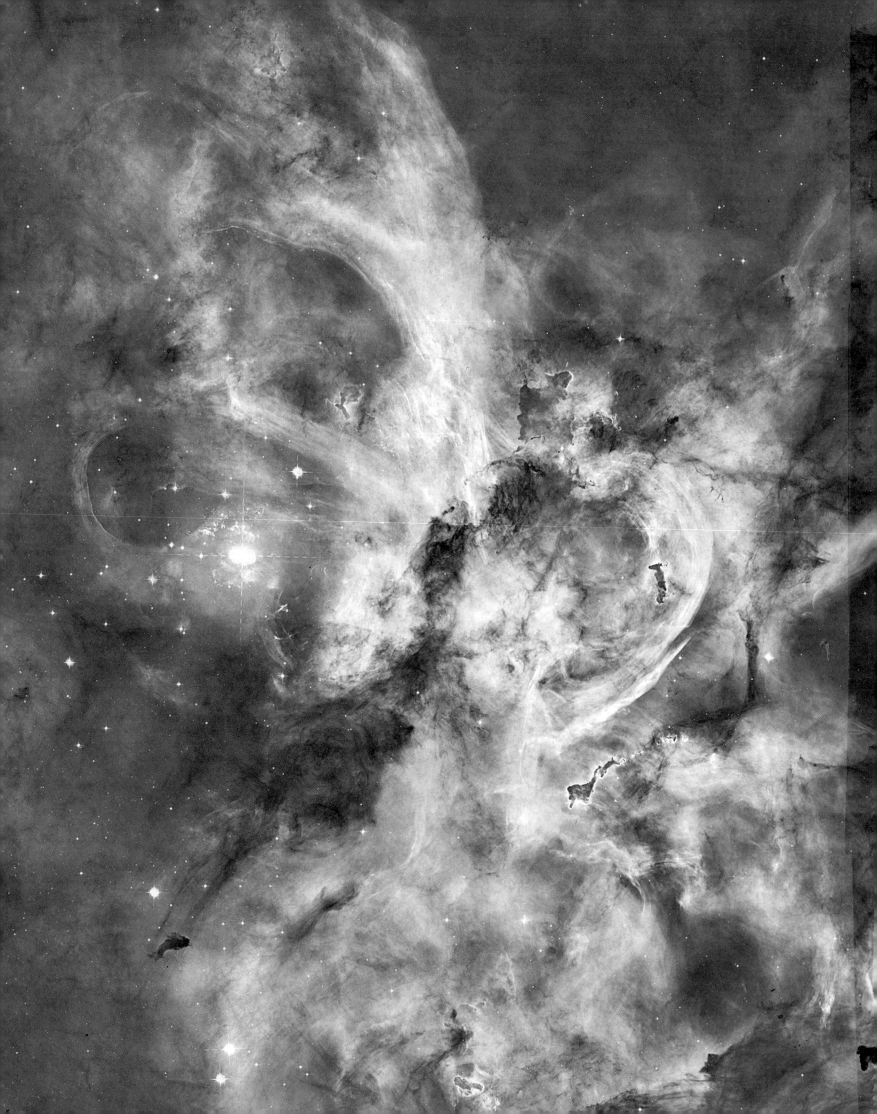